OPTICKS:

OR, A

TREATISE

OF THE

REFLEXIONS, REFRACTIONS, INFLEXIONS and COLOURS

OF

LIGHT.

ALSO

Two TREATISES

OF THE

SPECIES and MAGNITUDE

OF

Curvilinear Figures.

LONDON,

Printed for Sam. Smith, and Benj. Walford.
Printers to the Royal Society, at the *Prince's Arms* in
St. *Paul's* Church-yard. MDCCIV.

ADVERTISEMENT.

PArt of the enfuing Difcourfe about Light was written at the defire of fome Gentlemen of the Royal Society, in the Year 1675. and then fent to their Secretary, and read at their Meetings, and the reft was added about Twelve Years after to complete the Theory; except the Third Book, and the laft Propofition of the Second, which were fince put together out of fcattered Papers. To avoid being engaged in Difputes about thefe Matters, I have hitherto delayed the Printing, and fhould ftill have delayed it, had not the importunity of Friends prevailed upon me. If any other Papers writ on this Subject are got out of my Hands they are imperfect, and were perhaps written before I had tried all the Experiments here fet down, and fully fatisfied my felf about the Laws of Refractions and Compofition of Colours. I have here Publifhed what I think proper to come abroad, wifhing that it may not be Tranflated into another Language without my Confent.

The Crowns of Colours, which fometimes appear about the Sun and Moon, I have endeavoured to give an Account of; but for want of fufficient Obfervations leave that Matter to be further examined. The Subject of the Third Book I have alfo left imperfect, not having tried all the

Expe-

Experiments which I intended when I was about these Matters, nor repeated some of those which I did try, until I had satisfied my self about all their Circumstances. To communicate what I have tried, and leave the rest to others for further Enquiry, is all my Design in publishing these Papers.

In a Letter written to Mr. Leibnitz in the Year 1676. and published by Dr. Wallis, I mentioned a Method by which I had found some general Theorems about squaring Curvilinear Figures, or comparing them with the Conic Sections, or other the simplest Figures with which they may be compared. And some Years ago I lent out a Manuscript containing such Theorems, and having since met with some Things copied out of it, I have on this Occasion made it publick, prefixing to it an Introduction *and subjoyning a* Scholium *concerning that Method. And I have joined with it another small Tract concerning the Curvilinear Figures of the Second Kind, which was also written many Years ago, and made known to some Friends, who have solicited the making it publick.*

I. N.

The FIRST BOOK
OF
OPTICKS.

PART I.

MY Defign in this Book is not to explain the Properties of Light by Hypothefes, but to propofe and prove them by Reafon and Experiments: In order to which, I fhall premife the following Definitions and Axioms.

DEFINITIONS.

DEFIN. I.

BY the *Rays of Light I underftand its leaft Parts, and thofe as well Succeffive in the fame Lines as Contemporary in feveral Lines.* For it is manifeft that Light confifts of parts both Succeffive and Contemporary ; becaufe in the fame place you may ftop that which comes one moment, and let pafs that which comes prefently after; and in the fame time you may ftop it in any one place, and let it pafs in any other. For that part of Light which is ftopt cannot be the fame with that which is let pafs. The leaft Light or part of Light, which may be ftopt alone without the reft of the Light, or propagated alone, or do or fuffer any

A   thing

thing alone, which the reſt of the Light doth not or ſuf-
ers not, I call a Ray of Light.

DEFIN. II.

*Refrangibility of the Rays of Light, is their Diſpoſition to be
refracted or turned out of their Way in paſſing out of one tranſparent Body or Medium into another. And a greater or leſs Refrangibility of Rays, is their Diſpoſition to be turned more or leſs
out of their Way in like Incidences on the ſame Medium.* Mathematicians uſually conſider the Rays of Light to be Lines
reaching from the luminous Body to the body illuminated, and the refraction of thoſe Rays to be the bending
or breaking of thoſe Lines in their paſſing out of one Medium into another. And thus may Rays and Refractions
be conſidered, if Light be propagated in an inſtant. But
by an Argument taken from the Æquations of the times
of the Eclipſes of *Jupiter's Satellites* it ſeems that Light is
propagated in time, ſpending in its paſſage from the Sun
to us about Seven Minutes of time : And therefore I have
choſen to define Rays and Refractions in ſuch general
terms as may agree to Light in both caſes.

DEFIN. III.

*Reflexibility of Rays, is their Diſpoſition to be turned back into
the ſame Medium from any other Medium upon whoſe Surface they
fall. And Rays are more or leß reflexible, which are returned
back more or leſs eaſily.* As if Light paſs out of Glaſs into
Air, and by being inclined more and more to the common Surface of the Glaſs and Air, begins at length to be
totally reflected by that Surface; thoſe ſorts of Rays which
at like Incidences are reflected moſt copiouſly, or by inclining the Rays begin ſooneſt to be totally reflected, are
moſt reflexible.   D E-

DEFIN. IV.

The Angle of Incidence, is that Angle which the Line described by the incident Ray contains with the Perpendicular to the reflecting or refracting Surface at the Point of Incidence.

DEFIN. V.

The Angle of Reflexion or Refraction, is the Angle which the Line described by the reflected or refracted Ray containeth with the Perpendicular to the reflecting or refracting Surface at the Point of Incidence.

DEFIN. VI.

The Sines of Incidence, Reflexion, and Refraction, are the Sines of the Angles of Incidence, Reflexion, and Refraction.

DEFIN. VII.

The Light whose Rays are all alike Refrangible, I call Simple, Homogeneal and Similar; and that whose Rays are some more Refrangible than others, I call Compound, Heterogeneal and Dissimilar. The former Light I call Homogeneal, not because I would affirm it so in all respects; but because the Rays which agree in Refrangibility, agree at least in all those their other Properties. Which I consider in the following Discourse.

DEFIN. VIII.

The Colours of Homogeneal Lights, I call Primary, Homogeneal and Simple; and those of Heterogeneal Lights, Heterogeneal and Compound. For these are always compounded of the colours of Homogeneal Lights; as will appear in the following Discourse.　A 2　　　　　A XI-

A X I O M S.

A X. I.

THE *Angles of Incidence, Reflexion, and Refraction, lye in one and the same Plane.*

A X. II.

The Angle of Reflexion is equal to the Angle of Incidence.

A X. III.

If the refracted Ray be returned directly back to the Point of Incidence, it shall be refracted into the Line before described by the incident Ray.

A X. IV.

Refraction out of the rarer Medium into the denser, is made towards the Perpendicular ; that is, so that the Angle of Refraction be less than the Angle of Incidence.

A X. V.

The Sine of Incidence, is either accurately or very nearly in a given Ratio to the Sine of Refraction.

Whence if that Proportion be known in any one Inclination of the incident Ray, 'tis known in all the Inclinations, and thereby the Refraction in all cases of Incidence on the same refracting Body may be determined. Thus if the Refraction be made out of Air into Water, the Sine of Incidence of the red Light is to the Sine of its Refraction as 4 to 3. If out of Air into Glass, the Sines are

as

as 17 to 11. In Light of other Colours the Sines have other Proportions : but the difference is so little that it need seldom be considered.

Suppose therefore, that R S represents the Surface of *Fig.* 1. stagnating Water, and C is the point of Incidence in which any Ray coming in the Air from A in the Line A C is reflected or refracted, and I would know whether this Ray shall go after Reflexion or Refraction : I erect upon the Surface of the Water from the point of Incidence the Perpendicular C P and produce it downwards to Q, and conclude by the first Axiom, that the Ray after Reflexion and Refraction, shall be found somewhere in the Plane of the Angle of Incidence A C P produced. I let fall therefore upon the Perpendicular C P the Sine of Incidence A D, and if the reflected Ray be desired, I produce A D to B so that D B be equal to A D, and draw C B. For this Line C B shall be the reflected Ray; the Angle of Reflexion B C P and its Sine B D being equal to the Angle and Sine of Incidence, as they ought to be by the second Axiom. But if the refracted Ray be desired, I produce A D to H, so that D H may be to A D as the Sine of Refraction to the Sine of Incidence, that is as 3 to 4 ; and about the Center C and in the Plane A C P with the Radius C A describing a Circle A B E I draw Parallel to the Perpendicular C P Q, the Line H E cutting the circumference in E, and joyning C E, this Line C E shall be the Line of the refracted Ray. For if E F be let fall perpendicularly on the Line P Q, this Line E F shall be the Sine of Refraction of the Ray C E, the Angle of Refraction being E C Q; and this Sine E F is equal to D H, and consequently in Proportion to the Sine of Incidence A D as 3 to 4.

In

In like manner, if there be a Prism of Glass (that is a Glass bounded with two Equal and Parallel Triangular ends, and three plane and well polished Sides, which meet in three Parallel Lines running from the three Angles of one end to the three Angles of the other end) and if the Refraction of the Light in passing cross this Prism be desired : Let A C B represent a Plane cutting this Prism transversly to its three Parallel lines or edges there where the Light passeth through it, and let *d* E be the Ray incident upon the first side of the Prism A C where the Light goes into the Glass; And by putting the Proportion of the Sine of Incidence to the Sine of Refraction as 17 to 11 find E F the first refracted Ray. Then taking this Ray for the Incident Ray upon the second side of the Glass B C where the Light goes out, find the next refracted Ray F G by putting the Proportion of the Sine of Incidence to the Sine of Refraction as 11 to 17. For if the Sine of Incidence out of Air into Glass be to the Sine of Refraction as 17 to 11, the Sine of Incidence out of Glass into Air must on the contrary be to the Sine of Refraction as 11 to 17, by the third Axiom.

Fig. 2.

Fig. 3. Much after the same manner , if A C B D represent a Glass spherically Convex on both sides (usually called a *Lens,* such as is a Burning-glass, or Spectacle-glass, or an Object-glass of a Telescope) and it be required to know how Light falling upon it from any lucid point Q shall be refracted, let Q M represent a Ray falling upon any point M of its first spherical Surface A C B, and by erecting a Perpendicular to the Glass at the point M, find the first refracted Ray M N by the Proportion of the Sines 17 to 11. Let that Ray in going out of the Glass be incident upon N, and then find the second refracted Ray N *q* by the Proportion of the Sines 11 to 17. And after the same

same manner may the Refraction be found when the Lens is Convex on one side and Plane or Concave on the other, or Concave on both Sides.

A X. VI.

Homogeneal Rays which flow from several Points of any Object, and fall almost Perpendicularly on any reflecting or refracting Plane or Spherical Surface, shall afterwards diverge from so many other Points, or be Parallel to so many other Lines, or converge to so many other Points, either accurately or without any sensible Error. And the same thing will happen, if the Rays be reflected or refracted successively by two or three or more Plane or spherical Surfaces.

The Point from which Rays diverge or to which they converge may be called their *Focus*. And the Focus of the incident Rays being given, that of the reflected or refracted ones may be found by finding the Refraction of any two Rays, as above; or more readily thus.

Caf. 1. Let A C B be a reflecting or refracting Plane, *Fig.* 4. and Q the Focus of the incident Rays, and Q *q* C a perpendicular to that Plane. And if this perpendicular be produced to *q*, so that *q* C be equal to Q C, the point *q* shall be the Focus of the reflected Rays. Or if *q* C be taken on the same side of the Plane with Q C and in Proportion to Q C as the Sine of Incidence to the Sine of Refraction, the point *q* shall be the Focus of the refracted Rays.

Caf. 2. Let A C B be the reflecting Surface of any *Fig.* 5. Sphere whose Center is E. Bisect any Radius thereof (suppose E C) in T, and if in that Radius on the same side the point T you take the Points Q and *q*, so that T Q, T E, and T *q* be continual Proportionals, and the point Q be

the

the Focus of the incident Rays, the point *q* shall be the Focus of the reflected ones.

Fig. 6.   *Caf.* 3. Let A C B be the refracting Surface of any Sphere whose Center is E. In any Radius thereof E C produced both ways take E T and C *t* severally in such Proportion to that Radius as the lesser of the Sines of Incidence and Refraction hath to the difference of those Sines. And then if in the same Line you find any two Points Q and *q*, so that T Q be to E T as E *t* to *t q*, taking *t q* the contrary way from *t* which T Q lieth from T, and if the Point Q be the Focus of any incident Rays, the Point *q* shall be the Focus of the refracted ones.

And by the same means the Focus of the Rays after two or more Reflexions or Refractions may be found.

Fig. 7.   *Caf.* 4. Let A C B D be any refracting Lens, spherically Convex or Concave or Plane on either side, and let C D be its Axis (that is the Line which cuts both its Surfaces perpendicularly, and passes through the Centers of the Spheres,) and in this Axis let F and *f* be the Foci of the refracted Rays found as above, when the incident Rays on both sides the Lens are Parallel to the same Axis; and upon the Diameter F *f* bisected in E, describe a Circle. Suppose now that any Point Q be the Focus of any incident Rays. Draw Q E cutting the said Circle in T and *t*, and therein take *t q* in such Proportion to *t* E as *t* E or T E hath to T Q. Let *t q* lye the contrary way from *t* which T Q doth from T, and *q* shall be the Focus of the refracted Rays without any sensible Error, provided the Point Q be not so remote from the Axis, nor the Lens so broad as to make any of the Rays fall too obliquely on the refracting Surfaces.

And by the like Operations may the reflecting or refracting Surfaces be found when the two Foci are given,

and

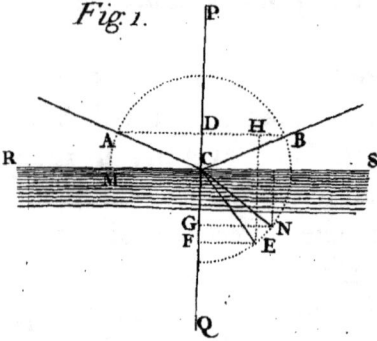

Fig. 1.

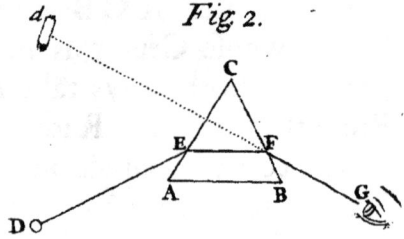

Fig. 2.

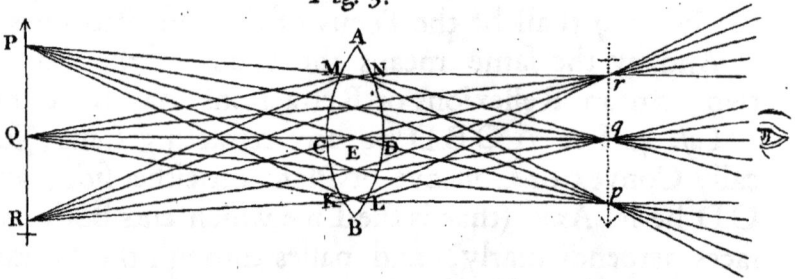

Fig. 3.

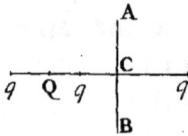

Fig. 4.

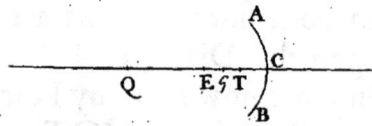

Fig. 5.

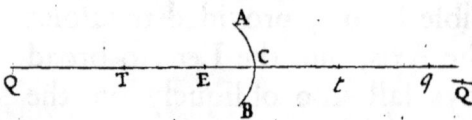

Fig. 6.

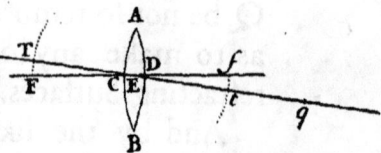

Fig. 7.

and thereby a Lens be formed, which shall make the Rays flow towards or from what place you please.

So then the meaning of this Axiom is, that if Rays fall upon any Plane or Spherical Surface or Lens, and before their Incidence flow from or towards any Point Q, they shall after Reflexion or Refraction flow from or towards the Point q found by the foregoing Rules. And if the incident Rays flow from or towards several points Q, the reflected or refracted Rays shall flow from or towards so many other Points q found by the same Rules. Whether the reflected and refracted Rays flow from or towards the Point q is easily known by the situation of that Point. For if that Point be on the same side of the reflecting or refracting Surface or Lens with the Point Q, and the incident Rays flow from the Point Q, the reflected flow towards the Point q and the refracted from it; and if the incident Rays flow towards Q, the reflected flow from q, and the refracted towards it. And the contrary happens when q is on the other side of that Surface.

A X. VII.

Wherever the Rays which come from all the Points of any Object meet again in so many Points after they have been made to converge by Reflexion or Refraction, there they will make a Picture of the Object upon any white Body on which they fall.

So if P R represent any Object without Doors, and A B *Fig.* 3. be a Lens placed at a hole in the Window-shut of a dark Chamber, whereby the Rays that come from any Point Q of that Object are made to converge and meet again in the Point q; and if a Sheet of white Paper be held at q for the Light there to fall upon it : the Picture of that Object P R will appear upon the Paper in its proper Shape

B   and

and Colours. For as the Light which comes from the Point Q goes to the Point *q*, so the Light which comes from other Points P and R of the Object, will go to so many other correspondent Points *p* and *r* (as is manifest by the sixth Axiom;) so that every Point of the Object shall illuminate a correspondent Point of the Picture, and thereby make a Picture like the Object in Shape and Colour, this only excepted that the Picture shall be inverted. And this is the reason of that Vulgar Experiment of casting the Species of Objects from abroad upon a Wall or Sheet of white Paper in a dark Room.

Fig. 8.   In like manner when a Man views any Object P Q R, the Light which comes from the several Points of the Object is so refracted by the transparent skins and humours of the Eye, (that is by the outward coat E F G called the *Tunica Cornea*, and by the crystalline humour A B which is beyond the Pupil *m k*) as to converge and meet again at so many Points in the bottom of the Eye, and there to paint the Picture of the Object upon that skin (called the *Tunica Retina*) with which the bottom of the Eye is covered. For Anatomists when they have taken off from the bottom of the Eye that outward and most thick Coat called the *Dura Mater*, can then see through the thinner Coats the Pictures of Objects lively painted thereon. And these Pictures propagated by Motion along the Fibres of the Optick Nerves into the Brain, are the cause of Vision. For accordingly as these Pictures are perfect or imperfect, the Object is seen perfectly or imperfectly. If the Eye be tinged with any colour (as in the Disease of the *Jaundise*) so as to tinge the Pictures in the bottom of the Eye with that Colour, then all Objects appear tinged with the same Colour. If the humours of the Eye by old Age decay, so as by shrinking to make the *Cornea* and Coat of the *Crystalline*

stalline

ftalline humour grow flatter than before, the Light will not be refracted enough, and for want of a fufficient Refraction will not converge to the bottom of the Eye but to fome place beyond it , and by confequence paint in the bottom of the Eye a confufed Picture, and according to the indiftinctnefs of this Picture the Object will appear confufed. This is the reafon of the decay of Sight in old Men, and fhews why their Sight is mended by Spectacles. For thofe Convex-glaffes fupply the defect of plumpnefs in the Eye, and by encreafing the Refraction make theRays converge fooner fo as to convene diftinctly at the bottom of the Eye if the Glafs have a due degree of convexity. And the contrary happens in fhort-fighted Men whofe Eyes are too plump. For the Refraction being now too great, the Rays converge and convene in the Eyes before they come at the bottom ; and therefore the Picture made in the bottom and the Vifion caufed thereby will not be diftinct, unlefs the Object be brought fo near the Eye as that the place where the converging Rays convene may be removed to the bottom, or that the plumpnefs of the Eye be taken off and the Refractions diminifhed by a Concave-glafs of a due degree of Concavity, or laftly that by Age the Eye grow flatter till it come to a due Figure : For fhort-fighted Men fee remote Objects beft in Old Age, and therefore they are accounted to have the moft lafting Eyes.

A X. VIII.

An Object feen by Reflexion or Refraction, appears in that place from whence the Rays after their laft Reflexion or Refraction diverge in falling on the Spectator's Eye.

If the Object A be feen by Reflexion of a Looking- *Fig. 9.* glafs *m n,* it fhall appear, not in it's proper place A, but

behind

behind the Glaſs at *a*, from whence any Rays A B, A C, A D, which flow from one and the ſame Point of the Object, do after their Reflexion made in the Points B, C, D, diverge in going from the Glaſs to E, F, G, where they are incident on the Spectator's Eyes. For theſe Rays do make the ſame Picture in the bottom of the Eyes as if they had come from the Object really placed at *a* without the interpoſition of the Looking-glaſs ; and all Viſion is made according to the place and ſhape of that Picture.

Fig. 2.　　In like manner the Object D ſeen through a Priſm appears not in its proper place D, but is thence tranſlated to ſome other place *d* ſituated in the laſt refracted Ray F G drawn backward from F to *d*.

Fig. 10.　　And ſo the Object Q ſeen through the Lens A B, appears at the place *q* from whence the Rays diverge in paſſing from the Lens to the Eye. Now it is to be noted, that the Image of the Object at *q* is ſo much bigger or leſſer than the Object it ſelf at Q, as the diſtance of the Image at *q* from the Lens A B is bigger or leſs than the diſtance of the Object at Q from the ſame Lens. And if the Object be ſeen through two or more ſuch Convex or Concave-glaſſes, every Glaſs ſhall make a new Image, and the Object ſhall appear in the place and of the bigneſs of the laſt Image. Which conſideration unfolds the Theory of Microſcopes and Teleſcopes. For that Theory conſiſts in almoſt nothing elſe than the deſcribing ſuch Glaſſes as ſhall make the laſt Image of any Object as diſtinct and large and luminous as it can conveniently be made.

　　I have now given in Axioms and their Explications the ſumm of what hath hitherto been treated of in Opticks. For what hath been generally agreed on I content my ſelf to aſſume under the notion of Principles, in order to what I have further to write. And this may ſuffice for an

Intro-

Introduction to Readers of quick Wit and good Underftanding not yet verfed in Opticks : Although thofe who
are already acquainted with this Science, and have
handled Glaffes, will more readily apprehend what followeth.

PROPOSITIONS.

P R O P. I.   Theor. I.

LIGHTS which differ in Colour, differ alfo in Degrees of Refrangibility.

The Proof by Experiments.

Exper. 1. I took a black oblong ftiff Paper terminated
by Parallel Sides, and with a Perpendicular right Line
drawn crofs from one Side to the other, diftinguifhed it
into two equal Parts. One of thefe Parts I painted with
a red Colour and the other with a blew. The Paper was
very black, and the Colours intenfe and thickly laid on,
that the Phænomenon might be more confpicuous. This
Paper I viewed through a Prifm of folid Glafs, whofe two
Sides through which the Light paffed to the Eye were
plane and well polifhed, and contained an Angle of about
Sixty Degrees : which Angle I call the refracting Angle of
the Prifm. And whilft I viewed it, I held it before a
Window in fuch manner that the Sides of the Paper were
parallel to the Prifm, and both thofe Sides and the Prifm
parallel to the Horizon, and the crofs Line perpendicular
to it ; and that the Light which fell from the Window
upon

upon the Paper made an Angle with the Paper, equal to that Angle which was made with the same Paper by the Light reflected from it to the Eye. Beyond the Prism was the Wall of the Chamber under the Window covered over with black Cloth, and the Cloth was involved in Darkness that no Light might be reflected from thence, which in passing by the edges of the Paper to the Eye, might mingle it self with the Light of the Paper and obscure the Phænomenon thereof. These things being thus ordered, I found that if the refracting Angle of the Prism be turned upwards, so that the Paper may seem to be lifted upwards by the Refraction, its blew half will be lifted higher by the Refraction than its red half. But if the refracting Angle of the Prism be turned downward, so that the Paper may seem to be carried lower by the Refraction, its blew half will be carried something lower thereby than its red half. Wherefore in both cases the Light which comes from the blew half of the Paper through the Prism to the Eye, does in like Circumstances suffer a greater Refraction than the Light which comes from the red half, and by consequence is more refrangible.

Fig. 11.   *Illustration.* In the Eleventh Figure, M N represents the Window, and D E the Paper terminated with parallel Sides D J and H E, and by the transverse Line F G distinguished into two halfs, the one D G of an intensely blew Colour, the other F E of an intensely red. And B A C c a b represents the Prism whose refracting Planes A B b a and A C c a meet in the edge of the refracting Angle A a. This edge A a being upward, is parallel both to the Horizon and to the parallel edges of the Paper D J and H E. And de represents the Image of the Paper seen by Refraction upwards in such manner that the blew half D G is carried higher to d g than the red half F E is to f e, and therefore
                                                        suffers

suffers a greater Refraction. If the edge of the refracting Angle be turned downward, the Image of the Paper will be refracted downward suppose to $\partial\epsilon$, and the blew half will be refracted lower to $\partial\gamma$ than the red half is to $\varphi\epsilon$.

Exper. 2. About the aforesaid Paper, whose two halfs were painted over with red and blew, and which was stiff like thin Pastboard, I lapped several times a slender thred of very black Silk, in such manner that the several parts of the thred might appear upon the Colours like so many black Lines drawn over them, or like long and slender dark Shadows cast upon them. I might have drawn black Lines with a Pen, but the threds were smaller and better defined. This Paper thus coloured and lined I set against a Wall perpendicularly to the Horizon, so that one of the Colours might stand to the right hand and the other to the left. Close before the Paper at the confine of the Colours below I placed a Candle to illuminate the Paper strongly : For the Experiment was tried in the Night. The flame of the Candle reached up to the lower edge of the Paper, or a very little higher. Then at the distance of Six Feet and one or two Inches from the Paper upon the Floor I erected a glass Lens four Inches and a quarter broad, which might collect the Rays coming from the several Points of the Paper, and make them converge towards so many other Points at the same distance of six Feet and one or two Inches on the other side of the Lens, and so form the Image of the coloured Paper upon a white Paper placed there; after the same manner that a Lens at a hole in a Window casts the Images of Objects abroad upon a Sheet of white Paper in a dark Room. The aforesaid white Paper, erected perpendicular to the Horizon and to the Rays which fell upon it from the Lens, I moved sometimes towards the Lens, sometimes from it, to find

the

the places where the Images of the blew and red parts of the coloured Paper appeared moſt diſtinct. Thoſe places I eaſily knew by the Images of the black Lines which I had made by winding the Silk about the Paper. For the Images of thoſe fine and ſlender Lines (which by reaſon of their blackneſs were like Shadows on the Colours) were confuſed and ſcarce viſible, unleſs when the Colours on either ſide of each Line were terminated moſt diſtinctly. Noting therefore, as diligently as I could, the places where the Images of the red and blew halfs of the coloured Paper appeared moſt diſtinct, I found that where the red half of the Paper appeared diſtinct, the blew half appeared confuſed, ſo that the black Lines drawn upon it could ſcarce be ſeen ; and on the contrary where the blew half appeared moſt diſtinct the red half appeared confuſed, ſo that the black Lines upon it were ſcarce viſible. And between the two places where theſe Images appeared diſtinct there was the diſtance of an Inch and a half : the diſtance of the white Paper from the Lens, when the Image of the red half of the coloured Paper appeared moſt diſtinct, being greater by an Inch and an half than the diſtance of the ſame white Paper from the Lens when the Image of the blew half appeared moſt diſtinct. In like Incidences therefore of the blew and red upon the Lens, the blew was refracted more by the Lens than the red, ſo as to converge ſooner by an Inch and an half, and therefore is more refrangible.

Fig. 1 2.   *Illuſtration.* In the Twelfth Figure, D E ſignifies the coloured Paper, D G the blew half, F E the red half, M N the Lens, H J the white Paper in that place where the red half with its black Lines appeared diſtinct, and *hi* the ſame Paper in that place where the blew half appeared diſtinct. The place *hi* was nearer to the Lens M N than the place H J by an Inch and an half.   *Scholium.*

Fig: 8.

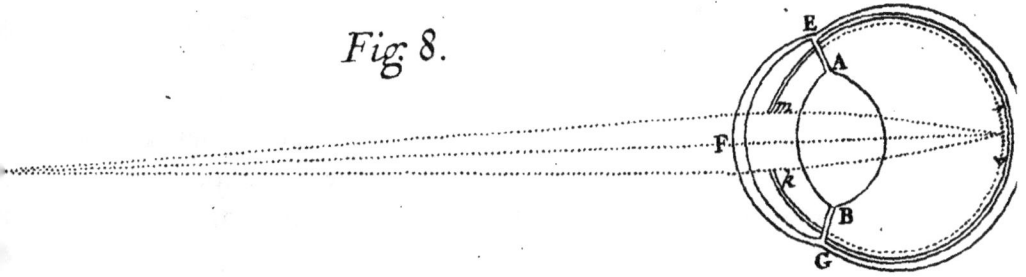

Fig: 9.

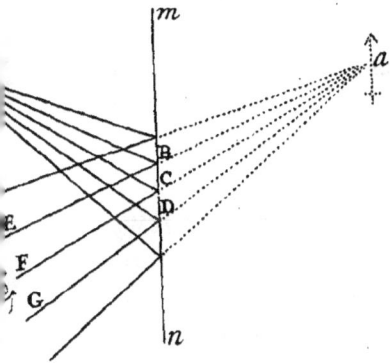

Fig: 11.

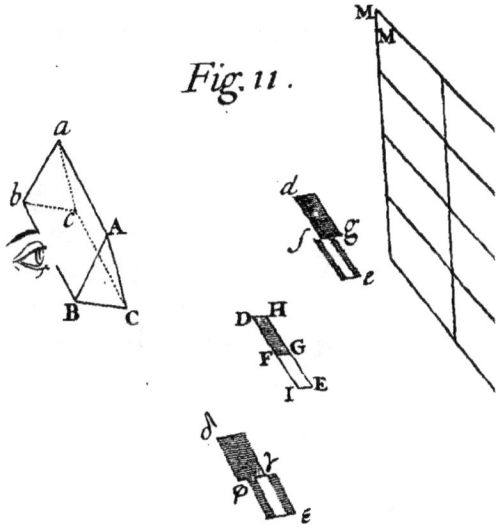

Fig 10.

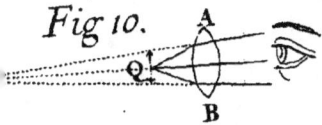

Fig: 12.

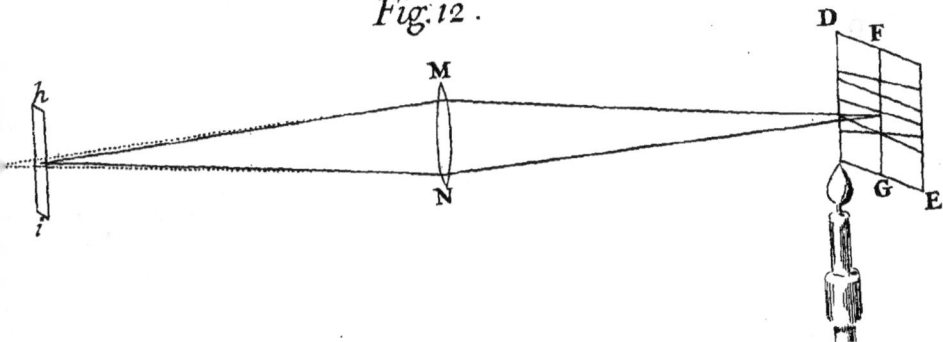

Scholium. The fame things fucceed notwithftanding that fome of the Circumftances be varied : as in the firft Experiment when the Prifm and Paper are any ways inclined to the Horizon, and in both when coloured Lines are drawn upon very black Paper. But in the Defcription of thefe Experiments, I have fet down fuch Circumftances by which either the Phænomenon might be rendred more confpicuous, or a Novice might more eafily try them, or by which I did try them only. The fame thing I have often done in the following Experiments : Concerning all which this one Admonition may fuffice. Now from thefe Experiments it follows not that all the Light of the blew is more Refrangible than all the Light of the red ; For both Lights are mixed of Rays differently Refrangible, So that in the red there are fome Rays not lefs Refrangible than thofe of the blew, and in the blew there are fome Rays not more Refrangible than thofe of the red ; But thefe Rays in Proportion to the whole Light are but few, and ferve to diminifh the Event of the Experiment, but are not able to deftroy it. For if the red and blew Colours were more dilute and weak, the diftance of the Images would be lefs than an Inch and an half ; and if they were more intenfe and full, that diftance would be greater, as will appear hereafter. Thefe Experiments may fuffice for the Colours of Natural Bodies. For in the Colours made by the Refraction of Prifms this Propofition will appear by the Experiments which are now to follow in the next Propofition.

C

PROP.

P R O P. II. Theor. II.

The Light of the Sun confiſts of Rays differently Refrangible.

The Proof by Experiments.

Exper. 3. IN a very dark Chamber at a round hole about one third part of an Inch broad made in the Shut of a Window I placed a Glaſs Priſm, whereby the beam of the Sun's Light which came in at that hole might be refracted upwards toward the oppoſite Wall of the Chamber, and there form a coloured Image of the Sun. The Axis of the Priſm (that is the Line paſſing through the middle of the Priſm from one end of it to the other end Parallel to the edge of the Refracting Angle) was in this and the following Experiments perpendicular to the incident Rays. About this Axis I turned the Priſm ſlowly, and ſaw the refracted Light on the Wall or coloured Image of the Sun firſt to deſcend and then to aſcend. Between the Deſcent and Aſcent when the Image ſeemed Stationary, I ſtopt the Priſm, and fixt it in that Poſture, that it ſhould be moved no more. For in that poſture the Refractions of the Light at the two ſides of the Refracting Angle, that is at the entrance of the Rays into the Priſm and at their going out of it, were equal to one another. So alſo in other Experiments as often as I would have the Refractions on both ſides the Priſm to be equal to one another, I noted the place where the Image of the Sun formed by the refracted Light ſtood ſtill between its two contrary Motions, in the common Period of its progreſs and egreſs; and when the Image fell upon that place, I made faſt the Priſm. And in this poſture, as

the

the moft convenient, it is to be underftood that all the Prifms are placed in the following Experiments, unlefs where fome other pofture is defcribed. The Prifm therefore being placed in this pofture, I let the refracted Light fall perpendicularly upon a Sheet of white Paper at the oppofite Wall of the Chamber, and obferved the Figure and Dimenfions of the Solar Image formed on the Paper by that Light. This Image was Oblong and not Oval, but terminated with two Rectilinear and Parallel Sides, and two Semicircular Ends. On its Sides it was bounded pretty diftinctly, but on its Ends very confufedly and indiftinctly, the Light there decaying and vanifhing by degrees. The breadth of this Image anfwered to the Sun's Diameter, and was about two Inches and the eighth part of an Inch, including the Penumbra. For the Image was eighteen Feet and an half diftant from the Prifm, and at this diftance that breadth if diminifhed by the Diameter of the hole in the Window-fhut, that is by a quarter of an Inch, fubtended an Angle at the Prifm of about half a Degree, which is the Sun's apparent Diameter. But the length of the Image was about ten Inches and a quarter, and the length of the Rectilinear Sides about eight Inches ; And the refracting Angle of the Prifm whereby fo great a length was made, was 64 degr. With a lefs Angle the length of the Image was lefs, the breadth remaining the fame. If the Prifm was turned about its Axis that way which made the Rays emerge more obliquely out of the fecond refracting Surface of the Prifm, the Image foon became an Inch or two longer, or more; and if the Prifm was turned about the contrary way, fo as to make the Rays fall more obliquely on the firft refracting Surface, the Image foon became an Inch or two fhorter. And therefore in trying this Experiment, I was as curious as I could be in placing the Prifm by the above-mentioned Rule exactly in

fuch

such a posture that the Refractions of the Rays at their emergence out of the Prism might be equal to that at their incidence on it. This Prism had some Veins running along within the Glass from one end to the other, which scattered some of the Sun's Light irregularly, but had no sensible effect in encreasing the length of the coloured Spectrum. For I tried the same Experiment with other Prisms with the same Success. And particularly with a Prism which seemed free from such Veins, and whose refracting Angle was $62\frac{1}{2}$ Degrees, I found the length of the Image $9\frac{3}{4}$ or 10 Inches at the distance of $18\frac{1}{2}$ Feet from the Prism, the breadth of the hole in the Window-shut being $\frac{1}{4}$ of an Inch as before. And because it is easie to commit a mistake in placing the Prism in its due posture, I repeated the Experiment four or five times, and always found the length of the Image that which is set down above. With another Prism of clearer Glass and better Pollish, which seemed free from Veins and whose refracting Angle was $63\frac{1}{2}$ Degrees, the length of this Image at the same distance of $18\frac{1}{2}$ Feet was also about 10 Inches, or $10\frac{1}{8}$. Beyond these Measures for about $\frac{1}{4}$ or $\frac{1}{3}$ of an Inch at either end of the Spectrum the Light of the Clouds seemed to be a little tinged with red and violet, but so very faintly that I suspected that tincture might either wholly or in great measure arise from some Rays of the Spectrum scattered irregularly by some inequalities in the Substance and Polish of the Glass, and therefore I did not include it in these Measures. Now the different Magnitude of the hole in the Window-shut, and different thickness of the Prism where the Rays passed through it, and different inclinations of the Prism to the Horizon, made no sensible changes in the length of the Image. Neither did the different matter of
the

the Prisms make any : for in a Vessel made of polished Plates of Glass cemented together in the shape of a Prism and filled with Water, there is the like Success of the Experiment according to the quantity of the Refraction. It is further to be observed, that the Rays went on in right Lines from the Prism to the Image, and therefore at their very going out of the Prism had all that Inclination to one another from which the length of the Image proceeded, that is the Inclination of more than two Degrees and an half. And yet according to the Laws of Opticks vulgarly received, they could not possibly be so much inclined to one another. For let E G represent the Window- *Fig.* 13. shut, F the hole made therein through which a beam of the Sun's Light was transmitted into the darkned Chamber, and A B C a Triangular Imaginary Plane whereby the Prism is feigned to be cut transversly through the middle of the Light. Or if you please, let A B C represent the Prism it self, looking directly towards the Spectator's Eye with its nearer end : And let X Y be the Sun, M N the Paper upon which the Solar Image or Spectrum is cast, and P T the Image it self whose sides tovvards V and W are Rectilinear and Parallel, and ends tovvards P and T Semicircular. Y K H P and X L J T are the two Rays, the first of which comes from the lower part of the Sun to the higher part of the Image, and is refracted in the Prism at K and H, and the latter comes from the higher part of the Sun to the lower part of the Image, and is refracted at L and J. Since the Refractions on both sides the Prism are equal to one another, that is the Refraction at K equal to the Refraction at J, and the Refraction at L equal to the Refraction at H, so that the Refractions of the incident Rays at K and L taken together are equal to the Refractions of the emergent Rays at H and J taken together :

ther : it follows by adding equal things to equal things, that the Refractions at K and H taken together, are equal to the Refractions at J and L taken together , and therefore the two Rays being equally refracted have the fame Inclination to one another after Refraction which they had before, that is the Inclination of half a Degree anfwering to the Sun's Diameter.   For fo great was the Inclination of the Rays to one another before Refraction.   So then, the length of the Image P T would by the Rules of Vulgar Opticks fubtend an Angle of half a Degree at the Prifm, and by confequence be equal to the breadth $v w$; and therefore the Image would be round.   Thus it would be were the two Rays X L J T and Y K H P and all the reft which form the Image P w T v, alike Refrangible. And therefore feeing by Experience it is found that the Image is not round but about five times longer than broad, the Rays which going to the upper end P of the Image fuffer the greateft Refraction, muft be more Refrangible than thofe which go to the lower end T , unlefs the inequality of Refraction be cafual.

This Image or Spectrum P T was coloured, being red at its leaft refracted end T, and violet at its moft refracted end P, and yellow green and blew in the intermediate fpaces. Which agrees with the firft Propofition, that Lights which differ in Colour do alfo differ in Refrangibility. The length of the Image in the foregoing Experiments I meafured from the fainteft and outmoft red at one end, to the fainteft and outmoft blew at the other end.

Exper. 4. In the Sun's beam which was propagated into the Room through the hole in the Window-fhut, at the diftance of fome Feet from the hole, I held the Prifm in fuch a pofture that its Axis might be perpendicular to that beam.   Then I looked through the Prifm upon the
hole,

hole, and turning the Prifm to and fro about its Axis to make the Image of the hole afcend and defcend, when between its two contrary Motions it feemed ftationary, I ftopt the Prifm that the Refractions on both fides of the refracting Angle might be equal to each other as in the former Experiment. In this Situation of the Prifm viewing through it the faid hole, I obferved the length of its refracted Image to be many times greater than its breadth, and that the moft refracted part thereof appeared violet, the leaft refracted red, the middle parts blew green and yellow in order. The fame thing happened when I removed the Prifm out of the Sun's Light, and looked through it upon the hole fhining by the Light of the Clouds beyond it. And yet if the Refraction were done regularly according to one certain Proportion of the Sines of Incidence and Refraction as is vulgarly fuppofed, the refracted Image ought to have appeared round.

So then, by thefe two Experiments it appears that in equal Incidences there is a confiderable inequality of Refractions : But whence this inequality arifes, whether it be that fome of the incident Rays are refracted more and others lefs, conftantly or by chance, or that one and the fame Ray is by Refraction difturbed, fhattered, dilated, and as it were fplit and fpread into many diverging Rays, as *Grimaldo* fuppofes, does not yet appear by thefe Experiments, but will appear by thofe that follow.

Exper. 5. Confidering therefore, that if in the third Experiment the Image of the Sun fhould be drawn out into an oblong form, either by a Dilatation of every Ray, or by any other cafual inequality of the Refractions, the fame oblong Image would by a fecond Refraction made Sideways be drawn out as much in breadth by the like Dilatation of the Rays or other cafual inequality of the Refractions

fractions Sideways, I tried what would be the Effects of such a second Refraction. For this end I ordered all things as in the third Experiment, and then placed a second Prism immediately after the first in a cross Position to it, that it might again refract the beam of the Sun's Light which came to it through the first Prism. In the first Prism this beam was refracted upwards, and in the second Sideways. And I found that by the Refraction of the second Prism the breadth of the Image was not increased, but its superior part which in the first Prism suffered the greater Refraction and appeared violet and blew, did again in the second Prism suffer a greater Refraction than its inferior part, which appeared red and yellow , and this without any Dilation of the Image in breadth.

Fig. 14.   *Illustration.*   Let S represent the Sun, F the hole in the Window, A B C the first Prism, D H the second Prism, Y the round Image of the Sun made by a direct beam of Light when the Prisms are taken away, P T the oblong Image of the Sun made by that beam passing through the first Prism alone when the second Prism is taken away, and *pt* the Image made by the cross Refractions of both Prisms together. Now if the Rays which tend towards the several Points of the round Image Y were dilated and spread by the Refraction of the first Prism, so that they should not any longer go in single Lines to single Points, but that every Ray being split, shattered, and changed from a Linear Ray to a Superficies of Rays diverging from the Point of Refraction, and lying in the Plane of the Angles of Incidence and Refraction, they should go in those Planes to so many Lines reaching almost from one end of the Image P T to the other, and if that Image should thence become oblong : those Rays and their several parts tending towards the several Points of

the

the Image P T ought to be again dilated and spread Sideways by the transverse Refraction of the second Prism, so as to compose a foursquare Image, such as is represented at *1. For the better understanding of which, let the Image P T be distinguished into five equal Parts PQK, KQRL, LRSM, MSVN, NVT. And by the same irregularity that the Orbicular Light Y is by the Refraction of the first Prism dilated and drawn out into a long Image P T, the the Light P Q K which takes up a space of the same length and breadth with the Light Y ought to be by the Refraction of the second Prism dilated and drawn out into the long Image *q kp, and the Light K Q R L into the long Image k q r l, and the Lights L R S M, M S V N, N V T into so many other long Images l r s m, m s v n, n v t 1; and all these long Images would compose the fourfquare Image *1. Thus it ought to be were every Ray dilated by Refraction, and spread into a triangular Superficies of Rays diverging from the Point of Refraction. For the second Refraction would spread the Rays one way as much as the first doth another, and so dilate the Image in breadth as much as the first doth in length. And the same thing ought to happen, were some Rays casually refracted more than others. But the Event is otherwise. The Image P T was not made broader by the Refraction of the second Prism, but only became oblique, as 'tis represented at p t, its upper end P being by the Refraction translated to a greater distance than its lower end T. So then the Light which went towards the upper end P of the Image, was (at equal Incidences) more refracted in the second Prism than the Light which tended towards the lower end T, that is the blew and violet, than the red and yellow ; and therefore was more Refrangible. The fame Light was by the Refraction of the first Prism translated further from the

D

place

place Y to which it tended before Refraction ; and there-fore suffered as well in the firſt Priſm as in the ſecond a greater Refraction than the reſt of the Light, and by con-ſequence was more Refrangible than the reſt, even before its incidence on the firſt Priſm.

Sometimes I placed a third Priſm after the ſecond, and ſometimes alſo a fourth after the third , by all which the Image might be often refracted ſideways : but the Rays which were more refracted than the reſt in the firſt Priſm were alſo more refracted in all the reſt, and that without any Dilatation of the Image ſideways : and therefore thoſe Rays for their conſtancy of a greater Refraction are de-ſervedly reputed more Refrangible.

Fig. 15. But that the meaning of this Experiment may more clearly appear, it is to be conſidered that the Rays which are equally Refrangible do fall upon a circle anſwering to the Sun's Diſque. For this was proved in the third Experi-ment. By a circle I underſtand not here a perfect Geo-metrical Circle, but any Orbicular Figure whoſe length is equal to its breadth, and which, as to ſenſe, may ſeem circular. Let therefore A G repreſent the circle which all the moſt Refrangible Rays propagated from the whole Diſque of the Sun, would illuminate and paint upon the oppoſite Wall if they were alone ; E L the circle which all the leaſt Refrangible Rays would in like manner illuminate and paint if they were alone ; B H, C J, D K, the circles which ſo many intermediate ſorts of Rays would ſuccef-ſively paint upon the Wall, if they were ſingly propagated from the Sun in ſucceſſive Order, the reſt being always in-tercepted ; And conceive that there are other intermediate Circles without number which innumerable other inter-mediate ſorts of Rays would ſucceſſively paint upon the Wall if the Sun ſhould ſucceſſively emit every ſort apart.

And

And feeing the Sun emits all thefe forts at once, they muft all together illuminate and paint innumerable equal circles, of all which, being according to their degrees of Refrangibility placed in order in a continual feries, that oblong Spectrum P T is compofed which I defcribed in the third Experiment. Now if the Sun's circular Image Y which is made by an unrefracted beam of Light was by any dilatation of the fingle Rays, or by any other irregularity in the Refraction of the firft Prifm, converted into the Oblong Spectrum, P T : then ought every circle A G, B H, C J, &c. in that Spectrum, by the crofs Refraction of the fecond Prifm again dilating or otherwife fcattering the Rays as before, to be in like manner drawn out and transformed into an Oblong Figure, and thereby the breadth of the Image P T would be now as much augmented as the length of the Image Y was before by the Refraction of the firft Prifm ; and thus by the Refractions of both Prifms together would be formed a fourfquare Figure $p \pi t \tau$ as I defcribed above. Wherefore fince the breadth of the Spectrum P T is not increafed by the Refraction fideways, it is certain that the Rays are not fplit or dilated, or otherways irregularly fcattered by that Refraction, but that every circle is by a regular and uniform Refraction tranflated entire into another place, as the circle A G by the greateft Refraction into the place ag, the circle B H by a lefs Refraction into the place bh, the circle C J by a Refraction ftill lefs into the place ci, and fo of the reft ; by which means a new Spectrum pt inclined to the former P T is in like manner compofed of circles lying in a right Line ; and thefe circles muft be of the fame bignefs with the former, becaufe the breadths of all the Spectrums Y, P T and pt at equal diftances from the Prifms are equal.

I con-

I confidered further that by the breadth of the hole F through which the Light enters into the Dark Chamber, there is a Penumbra made in the circuit of the Spectrum Y, and that Penumbra remains in the rectilinear Sides of the Spectrums P T and *pt*. I placed therefore at that hole a Lens or Object-glafs of a Telefcope which might caft the Image of the Sun diftinctly on Y without any Penumbra at all, and found that the Penumbra of the Rectilinear Sides of the oblong Spectrums P T and *pt* was alfo thereby taken away, fo that thofe Sides appeared as diftinctly defined as did the Circumference of the firft Image Y. Thus it happens if the Glafs of the Prifms be free from veins, atd their Sides be accurately plane and well polifhed without thofe numberlefs waves or curles which ufually arife from Sand-holes a little fmoothed in polifhing with Putty. If the Glafs be only well polifhed and free from veins and the Sides not accurately plane but a little Convex or Concave, as it frequently happens ; yet may the three Spectrums Y, P T and *pt* want Penumbras, but not in equal diftances from the Prifms. Now from this want of Penumbras, I knew more certainly that every one of the circles was refracted according to fome moft regular, uniform, and conftant law. For if there were any irregularity in the Refraction, the right Lines A E and G L which all the circles in the Spectrum P T do touch, could not by that Refraction be tranflated into the Lines *a e* and *g l* as diftinct and ftraight as they were before, but there would arife in thofe tranflated Lines fome Penumbra or crookednefs or undulation, or other fenfible Perturbation contrary to what is found by Experience. Whatfoever Penumbra or Perturbation fhould be made in the circles by the crofs Refraction of the fecond Prifm, all that Penumbra or Perturbation would be confpicuous in

the

the right Lines *a e* and *g l* which touch thofe circles. And therefore fince there is no fuch Penumbra or Perturbation in thofe right Lines there muft be none in the circles. Since the diftance between thofe Tangents or breadth of the Spectrum is not increafed by the Refractions, the Diameters of the circles are not increafed thereby. Since thofe Tangents continue to be right Lines, every circle which in the firft Prifm is more or lefs refracted, is exactly in the fame Proportion more or lefs refracted in the fecond. And feeing all thefe things continue to fucceed after the fame manner when the Rays are again in a third Prifm, and again in a fourth refracted Sideways, it is evident that the Rays of one and the fame circle as to their degree of Refrangibility continue always Uniform and Homogeneal to one another, and that thofe of feveral circles do differ in degree of Refrangibility, and that in fome certain and conftant Proportion. Which is the thing I was to prove.

There is yet another Circumftance or two of this Ex-*Fig.* 16. periment by which it becomes ftill more plain and convincing. Let the fecond Prifm D H be placed not immeately after after the firft, but at fome diftance from it ; Suppofe in the mid-way between it and the Wall on which the oblong Spectrum P T is caft, fo that the Light from the firft Prifm may fall upon it in the form of an oblong Spectrum, $\pi 1$ Parallel to this fecond Prifm, and be refracted Sideways to form the oblong Spectrum *p t* upon the Wall. And you will find as before, that this Spectrum *p t* is inclined to that Spectrum P T, which the firft Prifm forms alone without the fecond ; the blew ends P and *p* being further diftant from one another than the red ones T and *t*, and by confequence that the Rays which go to the blew end π of the Image $\pi 1$ and which therefore fuffer the greateft Refraction in the firft Prifm, are again in the fecond Prifm more refracted than the reft. The

Fig. 17. The fame thing I try'd alfo by letting the Sun's Light into a dark Room through two little round holes F and φ made in the Window, and with two Parallel Prifms A B C and α β γ placed at thofe holes (one at each) refracting thofe two beams of Light to the oppofite Wall of the Chamber, in fuch manner that the two colour'd Images P T and *m n* which they there painted were joyned end to end and lay in one ftraight Line, the red end T of the one touching the blew end *m* of the other. For if thefe two refracted beams were again by a third Prifm D H placed croft to the two firft, refracted Sideways, and the Spectrums thereby tranflated to fome other part of the Wall of the Chamber, fuppofe the Spectrum P T to *p t* and the Spectrum M N to *m n*, thefe tranflated Spectrums *p t* and *m n* would not lie in one ftraight Line with their ends contiguous as before, but be broken off from one another and become Parallel, the blew end of the Image *m n* being by a greater Refraction tranflated farther from its former place M T, than the red end *t* of the other Image *p t* from the fame place M T which puts the Propofition paft difpute. And this happens whether the third Prifm D H be placed immediately after the two firft or at a great diftance from them, fo that the Light refracted in the two firft Prifms be either white and circular, or coloured and oblong when it falls on the third.

Exper. 6. In the middle of two thin Boards I made round holes a third part of an Inch in Diameter, and in the Window-fhut a much broader hole, being made to let into my darkned Chamber a large beam of the Sun's Light ; I placed a Prifm behind the Shut in that beam to refract it towards the oppofite Wall, and clofe behind the Prifm I fixed one of the Boards, in fuch manner that the middle of the refracted Light might pafs through the hole

made

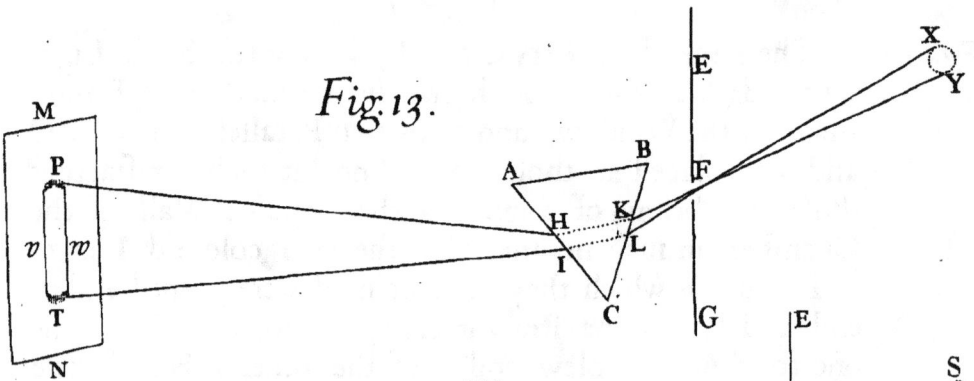

Fig. 13.

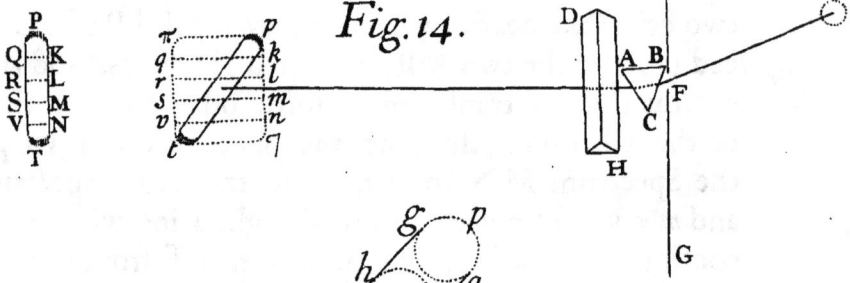

Fig. 14.

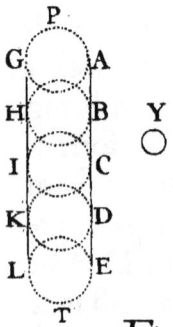

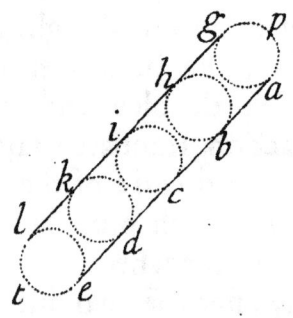

Fig. 15.

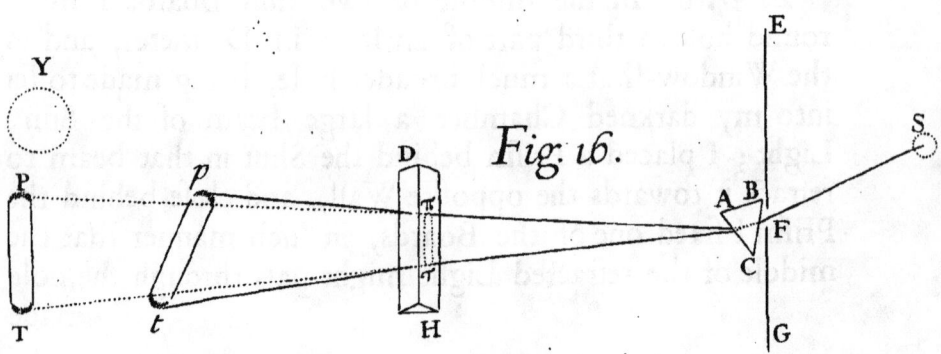

Fig. 16.

made in it, and the reſt be intercepted by the Board. Then at the diſtance of about twelve Feet from the firſt Board I fixed the other Board, in ſuch manner that the middle of the refracted Light which came through the hole in the firſt Board and fell upon the oppoſite Wall might paſs through the hole in this other Board, and the reſt being intercepted by the Board might paint upon it the coloured Spectrum of the Sun. And cloſe behind this Board I fixed another Priſm to refract the Light which came through the hole. Then I returned ſpeedily to the firſt Priſm, and by turning it ſlowly to and fro about its Axis, I cauſed the Image which fell upon the ſecond Board to move up and down upon that Board, that all its parts might ſucceſſively paſs through the hole in that Board and fall upon the Priſm behind it. And in the mean time, I noted the places on the oppoſite Wall to which that Light after its Refraction in the ſecond Priſm did paſs ; and by the difference of the places I found that the Light which being moſt refracted in the firſt Priſm did go to the blew end of the Image, was again more refracted in the ſecond Priſm than the Light which went to the red end of that Image, which proves as well the firſt Propoſition as the ſecond. And this happened whether the Axis of the two Priſms were parallel, or inclined to one another and to the Horizon in any given Angles.

Illuſtration. Let F be the wide hole in the Window-ſhut, *Fig.* 18. through which the Sun ſhines upon the firſt Priſm A B C, and let the refracted Light fall upon the middle of the Board D E, and the middle part of that Light upon the hole G made in the middle of that Board. Let this trajected part of the Light fall again upon the middle of the ſecond Board *d e* and there paint ſuch an oblong coloured Image of the Sun as was deſcribed in the third Experiment.

By

By turning the Prifm A B C flowly to and fro about its Axis this Image will be made to move up and down the Board *d e*, and by this means all its parts from one end to the other may be made to pafs fucceffively through the hole *g* which is made in the middle of that Board. In the mean while another Prifm *a b c* is to be fixed next after that hole *g* to refract the trajected Light a fecond time. And thefe things being thus ordered, I marked the places M and N of the oppofite Wall upon which the refracted Light fell, and found that whilft the two Boards and fecond Prifm remained unmoved, thofe places by turning the firft Prifm about its Axis were changed perpetually. For when the lower part of the Light which fell upon the fecond Board *d e* was caft through the hole *g* it went to a lower place M on the Wall, and when the higher part of that Light was caft through the fame hole *g*, it went to a higher place N on the Wall, and when any intermediate part of the Light was caft through that hole it went to fome place on the Wall between M and N. The unchanged Pofition of the holes in the Boards, made the Incidence of the Rays upon the fecond Prifm to be the fame in all cafes. And yet in that common Incidence fome of the Rays were more refracted and others lefs. And thofe were more refracted in this Prifm which by a greater Refraction in the firft Prifm were more turned out of the way, and therefore for their conftancy of being more refracted are defervedly called more Refrangible.

Exper. 7. At two holes made near one another in my Window-fhut I placed two Prifms, one at each, which might caft upon the oppofite Wall (after the manner of the third Experiment) two oblong coloured Images of the Sun. And at a little diftance from the Wall I placed a long flender Paper with ftraight and parallel edges, and

ordered

ordered the Prifms and Paper fo, that the red Colour of one Image might fall directly upon one half of the Paper, and the violet colour of the other Image upon the other half of the fame Paper; fo that the Paper appeared of two Colours, red and violet, much after the manner of the painted Paper in the firft and fecond Experiments. Then with a black Cloth I covered the Wall behind the Paper, that no Light might be reflected from it to difturb the Experiment, and viewing the Paper through a third Prifm held parallel to it, I faw that half of it which was illuminated by the Violet-light to be divided from the other half by a greater Refraction, efpecially when I went a good way off from the Paper. For when I viewed it too near at hand, the two halfs of the Paper did not appear fully divided from one another, but feemed contiguous at one of their Angles like the painted Paper in the firft Experiment. Which alfo happened when the Paper was too broad.

Sometimes inftead of the Paper I ufed a white Thred, and this appeared through the Prifm divided into two Parallel Threds as is reprefented in the 19th Figure, where *Fig.* 19. D G denotes the Thred illuminated with violet Light from D to E and with red Light from F to G, and *d e f g* are the parts of the Thred feen by Refraction. If one half of the Thred be conftantly illuminated with red, and the other half be illuminated with all the Colours fucceffively, (which may be done by caufing one of the Prifms to be turned about its Axis whilft the other remains unmoved) this other half in viewing the Thred through the Prifm, will appear in a continued right Line with the firft half when illuminated with red, and begin to be a little divided from it when illuminated with Orange, and remove further from it when illuminated with Yellow, and ftill

E

further

further when with Green, and further when with Blew, and
go yet further off when illuminated with Indigo, and fur-
theſt when with deep Violet. Which plainly ſhews, that
the Lights of ſeveral Colours are more and more Refrangible one than another, in this order of their Colours, Red,
Orange, Yellow, Green, Blew, Indigo, deep Violet ; and
ſo proves as well the firſt Propoſition as the ſecond.

Fig. 17.   I cauſed alſo the coloured Spectrums P T and M N
made in a dark Chamber by the Refractions of two Priſms
to lye in a right Line end to end, as was deſcribed above
in the fifth Experiment, and viewing them through a third
Priſm held Parallel to their length, they appeared no longer
in a right Line, but became broken from one another, as
they are repreſented at *p t* and *m n*, the violet end *m* of the
Spectrum *m n* being by a greater Refraction tranſlated
further from its former place M T than the red end *t* of the
other Spectrum *p t*.

Fig. 20.   I further cauſed thoſe two Spectrums P T and M N to
become co-incident in an inverted order of their Colours,
the red end of each falling on the violet end of the other,
as they are repreſented in the oblong Figure P T M N ;
and then viewing them through a Priſm D H held Parallel to their length, they appeared not co-incident as when
viewed with the naked Eye , but in the form of two diſtinct Spectrums *p t* and *m n* croſſing one another in the
middle after the manner of the letter X. Which ſhews
that the red of the one Spectrum and violet of the other,
which were co-incident at P N and M T , being parted
from one another by a greater Refraction of the violet to
p and *m* than of the red to *n* and *t*, do differ in degrees of
Refrangibility.

    I illuminated alſo a little circular piece of white Paper
all over with the Lights of both Priſms intermixed, and
when

when it was illuminated with the red of one Spectrum and deep violet of the other, so as by the mixture of those Colours to appear all over purple, I viewed the Paper, first at a less distance, and then at a greater, through a third Prism; and as I went from the Paper, the refracted Image thereof became more and more divided by the unequal Refraction of the two mixed Colours, and at length parted into two distinct Images, a red one and a violet one, whereof the violet was furthest from the Paper, and therefore suffered the greatest Refraction. And when that Prism at the Window which cast the violet on the Paper was taken away, the violet Image disappeared; but when the other Prism was taken away the red vanished : which shews that these two Images were nothing else than the Lights of the two Prisms which had been intermixed on the purple Paper, but were parted again by their unequal Refractions made in the third Prism through which the Paper was viewed. This also was observable that if one of the Prisms at the Window, suppose that which cast the violet on the Paper, was turned about its Axis to make all the Colours in this order, Violet, Indigo, Blew, Green, Yellow, Orange, Red, fall successively on the Paper from that Prism, the violet Image changed Colour accordingly, and in changing Colour came nearer to the red one, until when it was also red they both became fully co-incident.

I placed also two paper circles very near one another, the one in the red Light of one Prism, and the other in the violet Light of the other. The circles were each of them an Inch in Diameter, and behind them the Wall was dark that the Experiment might not be disturbed by any Light coming from thence. These circles thus illuminated, I viewed through a Prism so held that the Refraction might be made towards the red circle, and as I went from them

they

they came nearer and nearer together, and at length became co-incident; and afterwards when I went ſtill further off, they parted again in a contrary order, the violet by a greater Refraction being carried beyond the red.

Exper. 8. In Summer when the Sun's Light uſes to be ſtrongeſt, I placed a Priſm at the hole of the Window-ſhut, as in the third Experiment, yet ſo that its Axis might be Parallel to the Axis of the World, and at the oppoſite Wall in the Sun's refracted Light, I placed an open Book. Then going Six Feet and tvvo Inches from the Book, I placed there the abovementioned Lens, by vvhich the Light reflected from the Book might be made to converge and meet again at the diſtance of ſix Feet and tvvo Inches behind the Lens, and there paint the Species of the Book upon a ſheet of vvhite Paper much after the manner of the ſecond Experiment. The Book and Lens being made faſt, I noted the place vvhere the Paper vvas, vvhen the Letters of the Book, illuminated by the fulleſt red Light of the Solar Image falling upon it, did caſt their Species on that Paper moſt diſtinctly; And then I ſtay'd till by the Motion of the Sun and conſequent Motion of his Image on the Book, all the Colours from that red to the middle of the blew paſs'd over thoſe Letters; and when thoſe Letters were illuminated by that blew, I noted again the place of the Paper when they caſt their Species moſt diſtinctly upon it : And I found that this laſt place of the Paper was nearer to the Lens than its former place by about two Inches and an half, or two and three quarters. So much ſooner therefore did the Light in the violet end of the Image by a greater Refraction converge and meet, than the Light in the red end. But in trying this the Chamber was as dark as.I could make it. For if theſe Colours be diluted and weakned by the mixture of any adventitious Light, the diſtance

between

between the places of the Paper will not be so great. This distance in the second Experiment where the Colours of natural Bodies were made use of, was but an Inch and a half, by reason of the imperfection of those Colours. Here in the Colours of the Prism, which are manifestly more full, intense, and lively than those of natural Bodies, the distance is two Inches and three quarters. And were the Colours still more full, I question not but that the distance would be confiderably greater. For the coloured Light of the Prism, by the interfering of the Circles described in the 11th Figure of the fifth Experiment, and also by the Light of the very bright Clouds next the Sun's Body intermixing with these Colours, and by the Light scattered by the inequalities in the polish of the Prism, was so very much compounded, that the Species which those faint and dark Colours, the Indigo and Violet, cast upon the Paper were not distinct enough to be well observed.

Exper. 9. A Prism, whose two Angles at its Base were equal to one another and half right ones, and the third a right one, I placed in a beam of the Sun's Light let into a dark Chamber through a hole in the Window-shut as in the third Experiment. And turning the Prism slowly about its Axis until all the Light which went through one of its Angles and was refracted by it began to be reflected by its Base, at which till then it went out of the Glass, I observed that those Rays which had suffered the greatest Refraction were sooner reflected than the rest. I conceived therefore that those Rays of the reflected Light, which were most Refrangible, did first of all by a total Reflexion become more copious in that Light than the rest, and that afterwards the rest also, by a total Reflexion, became as copious as these. To try this, I made the reflected Light pass through another Prism, and being refracted

࿺ed by it to fall afterwards upon a sheet of white Paper placed at some distance behind it, and there by that Refraction to paint the usual Colours of the Prism. And then causing the first Prism to be turned about its Axis as above, I observed that when those Rays which in this Prism had suffered the greatest Refraction and appeared of a blew and violet Colour began to be totally reflected, the blew and violet Light on the Paper which was most refracted in the second Prism received a sensible increase above that of the red and yellow, which was least refracted; and afterwards when the rest of the Light which was green, yellow and red began to be totally reflected in the first Prism, the light of those Colours on the Paper received as great an increase as the violet and blew had done before. Whence 'tis manifest, that the beam of Light reflected by the Base of the Prism, being augmented first by the more Refrangible Rays and afterwards by the less Refrangible ones, is compounded of Rays differently Refrangible. And that all such reflected Light is of the same Nature with the Sun's Light, before its Incidence on the Base of the Prism, no Man ever doubted: it being generally allowed, that Light by such Reflexions suffers no Alteration in its Modifications and Properties. I do not here take notice of any Refractions made in the Sides of the first Prism, because the Light enters it perpendicularly at the first Side, and goes out perpendicularly at the second Side, and therefore suffers none. So then, the Sun's incident Light being of the same temper and constitution with his emergent Light, and the last being compounded of Rays differently Refrangible, the first must be in like manner compounded.

Fig. 21.   *Illustration.* In the 21th Figure, A B C is the first Prism, B C its Base, B and C its equal Angles at the Base, each

of

of 45 degrees, A its Rectangular Vertex, F M a beam of the Sun's Light let into a dark Room through a hole F one third part of an Inch broad, M its Incidence on the Base of the Prism, M G a less refracted Ray, M H a more refracted Ray, M N the beam of Light reflected from the Base, V X Y the second Prism by which this beam in passing through it is refracted, N *t* the less refracted Light of this beam, and N *p* the more refracted part thereof. When the first Prism A B C is turned about its Axis according to the order of the Letters A B C, the Rays M H emerge more and more obliquely out of that Prism, and at length after their most oblique Emergence are reflected towards N, and going on to *p* do increase the number of the Rays N *p*. Afterwards by continuing the motion of the first Prism, the Rays M G are also reflected to N and increase the number of the Rays N *t*. And therefore the Light M N admits into its Composition, first the more Refrangible Rays, and then the less Refrangible Rays, and yet after this Composition is of the same Nature with the Sun's immediate Light F M, the Reflexion of the specular Base B C causing no Alteration therein.

Exper. 10. Two Prisms, which were alike in shape, I tied so together, that their Axes and opposite Sides being Parallel, they composed a Parallelopiped. And, the Sun shining into my dark Chamber through a little hole in the Window-shut, I placed that Parallelopiped in his beam at some distance from the hole, in such a posture that the Axes of the Prisms might be perpendicular to the incident Rays, and that those Rays being incident upon the first Side of one Prism, might go on through the two contiguous Sides of both Prisms, and emerge out of the last Side of the second Prism. This Side being Parallel to the first Side of the first Prism, caused the emerging Light to be Parallel

to

to the Incident. Then, beyond thefe two Prifms I placed
a third, which might refract that emergent Light, and by
that Refraction caft the ufual Colours of the Prifm upon
the oppofite Wall, or upon a fheet of white Paper held at
a convenient diftance behind the Prifm for that refracted
Light to fall upon it. After this I turned the Parallelopiped
about its Axis, and found that when the contiguous Sides
of the two Prifms beeame fo oblique to the incident Rays
that thofe Rays began all of them to be reflected, thofe
Rays which in the third Prifm had fuffered the greateft Refraction and painted the Paper with violet and blew, were
firft of all by a total Reflexion taken out of the tranfmitted
Light, the reft remaining and on the Paper painting their
Colours of Green, Yellow, Orange, and Red as before ;
and afterwards by continuing the motion of the two Prifms,
the reft of the Rays alfo by a total Reflexion vanifhed in
order, according to their degrees of Refrangibility. The
Light therefore which emerged out of the two Prifms is
compounded of Rays differently Refrangible, feeing the
more Refrangible Rays may be taken out of it while the
lefs Refrangible remain. But this Light being trajected
only through the Parallel Superficies of the two Prifms, if
it fuffered any change by the Refraction of one Superficies
it loft that impreffion by the contrary Refraction of the
other Superficies, and fo being reftored to its priftine conftitution became of the fame nature and condition as at firft
before its Incidence on thofe Prifms; and therefore, before
its Incidence, was as much compounded of Rays differently
Refrangible as afterwards.

Fig. 22.   *Illuftration.* In the 22th Figure A B C and B C D are the
the two Prifms tied together in the form of a Parallelopiped, their Sides B C and C B being contiguous, and
their Sides A B and C D Parallel. And H J K is the third
Prifm,

Prifm, by which the Sun's Light propagated through the
hole F into the dark Chamber, and there paffing through
thofe fides of the Prifms AB, BC, CB and CD, is refracted at O to the white Paper P T, falling there partly upon
P by a greater Refraction, partly upon T by a lefs Refraction, and partly upon R and other intermediate places by
intermediate Refractions. By turning the Parallelopiped
ACBD about its Axis, according to the order of the Letters A, C, D, B, at length when the contiguous Planes BC
and CB become fufficiently oblique to the Rays FM,
which are incident upon them at M, there will vanifh totally out of the refracted Light OPT, firft of all the moft
refracted Rays OP, (the reft OR and OT remaining as
before) then the Rays OR and other intermediate ones,
and laftly, the leaft refracted Rays OT. For when the
Plane BC becomes fufficiently oblique to the Rays incident upon it, thofe Rays will begin to be totally reflected by it towards N; and firft the moft Refrangible Rays
will be totally reflected (as was explained in the preceding
experiment) and by confequence muft firft difappear at P,
and afterwards the reft as they are in order totally reflected to N, they muft difappear in the fame order at R and
T. So then the Rays which at O fuffer the greateft Refraction, may be taken out of the Light MO whilft the reft
of the Rays remain in it, and therefore that Light MO is
Compounded of Rays differently Refrangible. And becaufe the Planes AB and CD are parallel, and therefore
by equal and contrary Refractions deftroy one anothers
Effects, the incident Light FM muft be of the fame kind
and nature with the emergent Light MO, and therefore
doth alfo confift of Rays differently Refrangible. Thefe
two Lights FM and MO, before the moft refrangible Rays
are feparated out of the emergent Light MO agree in Colour,

F

lour, and in all other properties so far as my observation reaches, and therefore are deservedly reputed of the same Nature and Constitution, and by consequence the one is compounded as well as the other. But after the most Refrangible Rays begin to be totally reflected, and thereby separated out of the emergent Light MO, that Light changes its Colour from white to a dilute and faint yellow, a pretty good orange, a very full red successively and then totally vanishes. For after the most Refrangible Rays which paint the Paper at P with a Purple Colour, are by a total reflexion taken out of the Beam of light M O, the rest of the Colours which appear on the Paper at R and T being mixed in the light M O compound there a faint yellow, and after the blue and part of the green which appear on the Paper between P and R are taken away, the rest which appear between R and T (that is the Yellow, Orange, Red and a little Green) being mixed in the Beam M O compound there an Orange; and when all the Rays are by reflexion taken out of the Beam M O, except the least Refrangible, which at T appear of a full Red, their Colour is the same in that Beam M O as afterwards at T, the Refraction of the Prism H J K serving only to separate the differently Refrangible Rays, without making any alteration in their Colours, as shall be more fully proved hereafter. All which confirms as well the first Proposition as the second.

Scholium. If this Experiment and the former be conjoyned Fig. 22. and made one, by applying a fourth Prism V X Y to refract the reflected Beam M N towards *tp*, the conclusion will be clearer. For then the light N*p* which in the 4th Prism is more refracted, will become fuller and stronger when the Light O P, which in the third Prism H J K is more refracted, vanishes at P; and afterwards when the less refracted

refracted Light O T vaniſhes at T, the leſs refracted Light N*t* will become encreaſed whilſt the more refracted Light at *p* receives no further encreaſe. And as the trajected Beam M O in vaniſhing is always of ſuch a Colour as ought to reſult from the mixture of the Colours which fall upon the Paper P T, ſo is the reflected Beam M N always of ſuch a Colour as ought to reſult from the mixture of the Colours which fall upon the Paper *pt*. For when the moſt refrangible Rays are by a total Reflexion taken out of the Beam M O, and leave that Beam of an Orange Colour, the exceſs of thoſe Rays in the reflected Light, does not only make the Violet, Indigo and Blue at *p* more full, but alſo makes the Beam M N change from the yellowiſh Colour of the Sun's Light, to a pale white inclining to blue, and afterward recover its yellowiſh Colour again, ſo ſoon as all the reſt of the tranſmitted light M O T is reflected.

Now ſeeing that in all this variety of Experiments, whether the trial be made in Light reflected, and that either from natural Bodies, as in the firſt and ſecond Experiment, or Specular, as in the Ninth ; or in Light refracted, and that either before the unequally refracted Rays are by diverging ſeparated from one another, and loſing their whiteneſs which they have altogether, appear ſeverally of ſeveral Colours, as in the fifth Experiment ; or after they are ſeparated from one another, and appear Coloured as in the ſixth, ſeventh, and eighth Experiments ; or in Light trajected through Parallel ſuperficies, deſtroying each others Effects as in the 10th Experiment ; there are always found Rays, which at equal Incidences on the ſame Medium ſuffer unequal Refractions, and that without any ſplitting or dilating of ſingle Rays, or contingence in the inequality of the Refractions, as is proved in the fifth and ſixth Ex-

periments ;

periments; and feeing the Rays which differ in Refrangibility may be parted and forted from one another, and that
either by Refraction as in the third Experiment, or by Reflexion as in the tenth, and then the feveral forts apart at
equal Incidences fuffer unequal Refractions, and thofe forts
are more refracted than others after feparation, which were
more refracted before it, as in the fixth and following Experiments, and if the Sun's Light be trajected through three
or more crofs Prifms fucceffively, thofe Rays which in the
firft Prifm are refracted more than others are in all the following Prifms, refracted more then others in the fame rate
and proportion, as appears by the fifth Experiment; it's
manifeft that the Sun's Light is an Heterogeneous mixture of
Rays, fome of which are conftantly more Refrangible then
others, as was to be propofed.

P R O P. III.   Theor. III.

*The Sun's Light confifts of Rays differing in Reflexibility, and
thofe Rays are more Reflexible than others which are more Refrangible.*

THIS is manifeft by the ninth and tenth Experiments: For in the ninth Experiment, by turning
the Prifm about its Axis, until the Rays within it which in
going out into the Air were refracted by its Bafe, became
fo oblique to that Bafe, as to begin to be totally reflected
thereby; thofe Rays became firft of all totally reflected,
which before at equal Incidences with the reft had fuffered
the greateft Refraction. And the fame thing happens in
the Reflexion made by the common Bafe of the two Prifms
in the tenth Experiment.

P R O P.

Fig. 17.

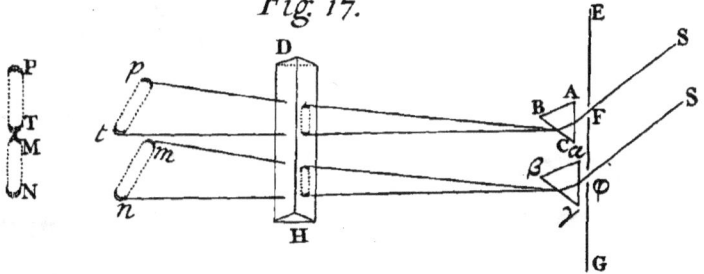

Fig. 18.

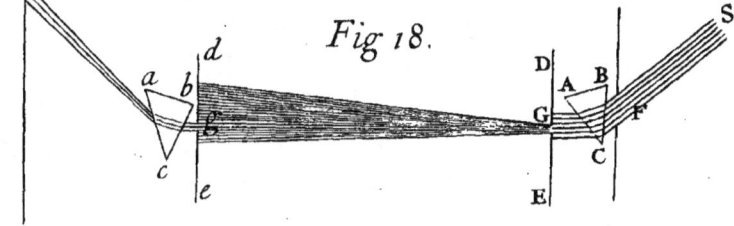

Fig. 19.

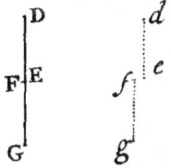

Fig. 20.

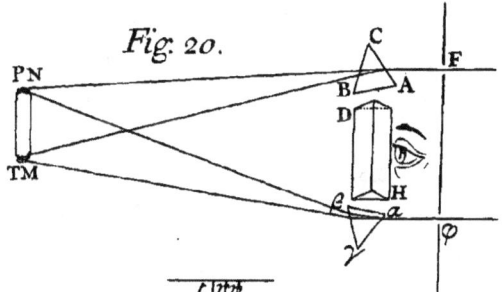

Fig. 21.

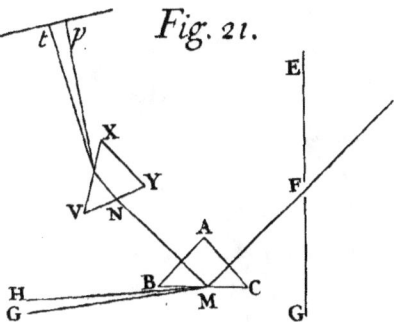

Fig. 22.

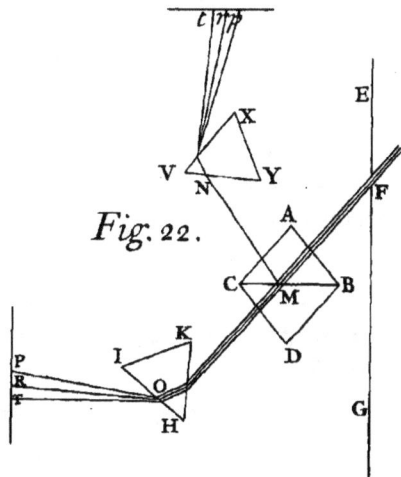

PROP. IV. Prob. I.

*To separate from one another the Heterogeneous Rays of
Compound Light.*

THE Heterogeneous Rays are in some measure separated from one another by the Refraction of the
Prism in the third Experiment, and in the fifth Experiment
by taking away the Penumbra from the Rectilinear sides of
the Coloured Image, that separation in those very Rectilinear sides or straight edges of the Image becomes perfect.
But in all places between those rectilinear edges, those innumerable Circles there described, which are severally illuminated by Homogeneral Rays, by interfering with one
another, and being every where commixt, do render the
Light sufficiently Compound. But if these Circles, whilst
their Centers keep their distances and positions, could be
made less in Diameter, their interfering one with another
and by consequence the mixture of the Heterogeneous
Rays would be proportionally diminished. In the 23th *Fig.* 23.
Figure let A G, B H, C J, D K, E L, F M be the Circles
which so many sorts of Rays flowing from the same Disque
of the Sun, do in the third Experiment illuminate ; of all
which and innumerable other intermediate ones lying in a
continual Series between the two Rectilinear and Parallel
edges of the Sun's oblong Image P T, that Image is composed as was explained in the fifth Experiment. And let
a g, *b h*, *c i*, *d k*, *e l*, *f m* be so many less Circles lying in
a like continual Series between two Parallel right Lines *a f*
and *g m* with the same distances between their Centers,
and illuminated by the same sorts of Rays, that is the
Circle *a g* with the same sort by which the corresponding
Circle

Circle A G was illuminated, and the Circle *bh* with the fam e
fort by which the corresponding Circle B H was illuminated,
and the reft of the Circles *c i*, *d k*, *e l*, *f m* refpectively,
with the fame forts of Rays by which the feveral corre-
fponding Circles C J, D K, E L, F M were illuminated.
In the Figure P T compofed of the greater Circles, three
of thofe Circles A G, B H, C J, are fo expanded into one
another, that the three forts of Rays by which thofe Circles are illuminated, together with other innumerable forts
of intermediate Rays, are mixed at Q R in the middle of
the Circle B H. And the like mixture happens through-
out almoft the whole length of the Figure P T. But in
the Figure *p t* compofed of the lefs Circles, the three lefs
Circles *a g*, *b h*, *c i*, which anfwer to thofe three greater, do
not extend into one another; nor are there any where
mingled fo much as any two of the three forts of Rays
by which thofe Circles are illuminated, and which in the
Figure P T are all of them intermingled at B H.

Now he that fhall thus confider it, will eafily underftand
that the mixture is diminifhed in the fame Proportion
with the Diameters of the Circles. If the Diameters of
the Circles whilft their Centers remain the fame, be made
three times lefs than before, the mixture will be alfo three
times lefs; if ten times lefs, the mixture will be ten times
lefs, and fo of other Proportions. That is, the mixture
of the Rays in the greater Figure P T will be to their mixture in the lefs *p t*, as the Latitude of the greater Figure is
to the Latitude of the lefs. For the Latitudes of thefe Figures are equal to the Diameters of their Circles. And
hence it eafily follows, that the mixture of the Rays in the
refracted Spectrum *p t* is to the mixture of the Rays in the
direct and immediate Light of the Sun, as the breadth of
that Spectrum is to the difference between the length and
breadth of the fame Spectrum.   So

So then, if we would diminish the mixture of the Rays, we are to diminish the Diameters of the Circles. Now these would be diminished if the Sun's Diameter to which they answer could be made less than it is, or (which comes to the same purpose) if without Doors, at a great distance from the Prism towards the Sun, some opake body were placed, with a round hole in the middle of it, to intercept all the Sun's Light, excepting so much as coming from the middle of his Body could pass through that hole to the Prism. For so the Circles A G, B H and the rest, would not any longer answer to the whole Disque of the Sun , but only to that part of it which could be seen from the Prism through that hole, that is to the apparent magnitude of that hole viewed from the Prism. But that these Circles may answer more distinctly to that hole a Lens is to be placed by the Prism to cast the Image of the hole, (that is, every one of the Circles A G, B H, &c.) distinctly upon the Paper at P T, after such a manner as by a Lens placed at a Window the Species of Objects abroad are cast distinctly upon a Paper within the Room, and the Rectilinear Sides of the oblong solar Image in the fifth Experiment became distinct without any Penumbra. If this be done it will not be necessary to place that hole very far off, no not beyond the Window. And therefore instead of that hole, I used the hole in the Window-shut as follows.

Exper. 11. In the Sun's Light let into my darkned Chamber through a small round hole in my Window-shut, at about 10 or 12 Feet from the Window, I placed a Lens, by which the Image of the hole might be distinctly cast upon a sheet of white Paper, placed at the distance of six, eight, ten or twelve Feet from the Lens. For according to the difference of the Lenses I used various

distances,

diftances, which I think not worth the while to defcribe. Then immediately after the Lens I placed a Prifm, by which the trajected Light might be refracted either upwards or fideways, and thereby the round Image which the Lens alone did caft upon the Paper might be drawn out into a long one with Parallel Sides, as in the third Experiment. This oblong Image I let fall upon another Paper at about the fame diftance from the Prifm as before, moving the Paper either towards the Prifm or from it, until I found the juft diftance where the Rectilinear Sides of the Image became moft diftinct. For in this cafe the circular Images of the hole which compofe that Image after the fame manner that the Circles *a g*, *b h*, *c i*, &c. do *Fig.* 23. the Figure *p t*, were terminated moft diftinctly without any Penumbra, and therefore extended into one another the leaft that they could, and by confequence the mixture of the Heterogeneous Rays was now the leaft of all. By this *Fig.* 23, means I ufed to form an oblong Image (fuch as is *p t*) of *and* 24. circular Images of the hole (fuch as are *a g*, *b h*, *c i*, &c.) and by ufing a greater or lefs hole in the Window-fhut, I made the circular Images *a g*, *b h*, *c i*, &c. of which it was formed, to become greater or lefs at pleafure, and thereby the mixture of the Rays in the Image *p t* to be as much or as little as I defired.

Fig. 24.   *Illuftration.* In the 24th Figure, F reprefents the circular hole in the Window-fhut, M N the Lens whereby the Image or Species of that hole is caft diftinctly upon a Paper at J, A B C the Prifm whereby the Rays are at their emerging out of the Lens refracted from J towards another Paper at *p t*, and the round Image at J is turned into an oblong Image *p t* falling on that other Paper. This Image *p t* confifts of Circles placed one after another in a Rectilinear order, as was fufficiently explained in the fifth

Experiment;

Experiment; and these Circles are equal to the Circle I, and consequently answer in Magnitude to the hole F; and therefore by diminishing that hole they may be at pleasure diminished, whil'st their Centers remain in their places. By this means I made the breadth of the Image pt to be forty times, and sometimes sixty or seventy times less than its length. As for instance, if the breadth of the hole F be $\frac{1}{10}$ of an Inch, and M F the distance of the Lens from the hole be 12 Feet; and if p B or p M the distance of the Image pt from the Prism or Lens be 10 Feet, and the refracting Angle of the Prism be 62 degrees, the breadth of the Image pt will be $\frac{1}{12}$ of an Inch and the length about six Inches, and therefore the length to the breadth as 72 to 1, and by consequence the Light of this Image 71 times less compound than the Sun's direct Light. And Light thus far Simple and Homogeneal, is sufficient for trying all the Experiments in this Book about simple Light. For the composition of Heterogeneal Rays is in this Light so little that it is scarce to be discovered and perceived by sense, except perhaps in the Indigo and Violet; for these being dark Colours, do easily suffer a sensible allay by that little scattering Light which uses to be refracted irregularly by the inequaliteis of the Prism.

Yet instead of the circular hole F, 'tis better to substitute an oblong hole shaped like a long Parallelogram with its length Parallel to the Prism A B C. For if this hole be an Inch or two long, and but a tenth or twentieth part of an Inch broad or narrower : the Light of the Image pt will be as Simple as before or simpler, and the Image will become much broader, and therefore more fit to have Experiments tried in its Light than before.

Instead of this Parallelogram-hole may be substituted a Triangular one of equal Sides, whose Base for instance is

G   abou:

about the tenth part of an Inch, and its height an Inch or more. For by this means, if the Axis of the Prism be Parallel to the Perpendicular of the Triangle, the Image *Fig.* 25. *p t* will now be formed of Equicrural Triangles *a g, b h, c i, d k, e l, f m,* &c. and innumerable other intermediate ones answering to the Triangular hole in shape and bigness, and lying one after another in a continual Series between two Parallel Lines *a f* and *g m.* These Triangles are a little intermingled at their Bases but not at their Vertices, and therefore the Light on the brighter side *a f* of the Image where the Bases of the Triangles are is a little compounded, but on the darker side *g m* is altogether uncompounded, and in all places between the sides the Composition is Proportional to the distances of the places from that obscurer side *g m.* And having a Spectrum *p t* of such a Composition, we may try Experiments either in its stronger and less simple Light near the side *a f,* or in its weaker and simpler Light near the other side *l m,* as it shall seem most convenient.

But in making Experiments of this kind the Chamber ought to be made as dark as can be, least any forreign Light mingle it self with the Light of the Spectrum *p t,* and render it compound; especially if we would try Experiments in the more simple Light next the side *g m* of the Spectrum; which being fainter, will have a less Proportion to the forreign Light, and so by the mixture of that Light be more troubled and made more compound. The Lens also ought to be good, such as may serve for Optical Uses, and the Prism ought to have a large Angle, suppose of 70 degrees, and to be well wrought, being made of Glass free from Bubbles and Veins, with its sides not a little Convex or Concave as usually happens but truly Plane, and its pollish elaborate, as in working Optick-

glasses,

glaffes, and not fuch as is ufually wrought with Putty, whereby the edges of the Sand-holes being worn away, there are left all over the Glafs a numberlefs company of very little Convex polite rifings like Waves. The edges alfo of the Prifm and Lens fo far as they may make any irregular Refraction, muft be covered with a black Paper glewed on. And all the Light of the Sun's beam let into the Chamber which is ufelefs and unprofitable to the Experiment, ought to be intercepted with black Paper or other black Obftacles. For otherwife the ufelefs Light being reflected every way in the Chamber, will mix with the oblong Spectrum and help to difturb it. In trying thefe things fo much Diligence is not altogether neceffary, but it will promote the fuccefs of the Experiments, and by a very fcrupulous Examiner of things deferves to be applied. It's difficult to get glafs Prifms fit for this purpofe, and and therefore I ufed fometimes Prifmatick Veffels made with pieces of broken Looking-glaffes, and filled with rain Water. And to increafe the Refraction, I fometimes impregnated the Water ftrongly with *Saccharum Saturni.*

PROP. V. Theor. IV.

Homogeneal Light is refracted regularly without any Dilatation fplitting or fhattering of the Rays, and the confufed Vifion of Objects feen through Refracting Bodies by Heterogeneal Light arifes from the different Refrangibility of feveral forts of Rays.

THE firft Part of this Propofition has been already fufficiently proved in the fifth Experiment, and will further appear by the Experiments which follow.

Exper. 12.

Exper. 12. In the middle of a black Paper I made a round hole about a fifth or sixth part of an Inch in Diameter. Upon this Paper I caused the Spectrum of Homogeneal Light described in the former Proposition, so to fall, that some part of the Light might pass through the hole of the Paper. This transmitted part of the Light I refracted with a Prism placed behind the Paper, and letting this refracted Light fall perpendicularly upon a white Paper two or three Feet distant from the Prism, I found that the Spectrum formed on the Paper by this Light was not oblong, as when 'tis made (in the third Experiment) by Refracting the Sun's compound Light, but was (so far as I could judge by my Eye) perfectly circular, the length being no greater than the breadth. Which shews that this Light is refracted regularly without any Dilatation of the Rays.

Exper. 13. In the Homogeneal Light I placed a Circle of $\frac{1}{4}$ of an Inch in Diameter, and in the Sun's unrefracted Heterogeneal white Light I placed another Paper Circle of the same bigness. And going from the Papers to the distance of some Feet, I viewed both Circles through a Prism. The Circle illuminated by the Sun's Heterogeneal Light appeared very oblong as in the fourth Experiment, the length being many times greater than the breadth : but the other Circle illuminated with Homogeneal Light appeared Circular and distinctly defined as when 'tis viewed with the naked Eye. Which proves the whole Proposition.

Exper. 14. In the Homogeneal Light I placed Flies and such like Minute Objects, and viewing them through a Prism, I saw their Parts as distinctly defined as if I had viewed them with the naked Eye. The same Objects placed in the Sun's unrefracted Heterogeneal Light which was white I viewed also through a Prism, and saw them most

confusedly

confusedly defined, so that I could not distinguish their smaller Parts from one another. I placed also the Letters of a small Print one while in the Homogeneal Light and then in the Heterogeneal, and viewing them through a Prism, they appeared in the latter case so confused and indistinct that I could not read them ; but in the former they appeared so distinct that I could read readily, and thought I saw them as distinct as when I viewed them with my naked Eye. In both cases I viewed the same Objects through the same Prism at the same distance from me and in the same Situation. There was no difference but in the Light by which the Objects were illuminated, and which in one case was Simple and in the other Compound, and therefore the distinct Vision in the former case and confused in the latter could arise from nothing else than from that difference of the Lights. Which proves the whole Proposition.

And in these three Experiments it is further very remarkable, that the Colour of Homogeneal Light was never changed by the Refraction.

PROP. VI. Theor. V.

The Sine of Incidence of every Ray considered apart, is to its Sine of Refraction in a given Ratio.

THAT every Ray considered apart is constant to it self in some certain degree of Refrangibility, is sufficiently manifest out of what has been said. Those Rays which in the first Refraction are at equal Incidences most refracted, are also in the following Refractions at equal Incidences most refracted ; and so of the least Refrangible, and the rest which have any mean degree of
Refran-

Refrangibility, as is manifeſt by the 5th, 6th, 7th, 8th, and 9th Experiments. And thoſe which the firſt time at like Incidences are equally refracted, are again at like Incidences equally and uniformly refracted, and that whether they be refracted before they be ſeparated from one another as in the 5th Experiment, or whether they be refracted apart, as in the 12th, 13th and 14th Experiments. The Refraction therefore of every Ray apart is regular, and what Rule that Refraction obſerves we are now to ſhew.

The late Writers in Opticks teach, that the Sines of Incidence are in a given Proportion to the Sines of Refraction, as was explained in the 5th Axiom ; and ſome by Inſtruments fitted for meaſuring Refractions, or otherwiſe experimentally examining this Proportion, do acquaint us that they have found it accurate. But whilſt they, not underſtanding the different Refrangibility of ſeveral Rays, conceived them all to be refracted according to one and the ſame Proportion, 'tis to be preſumed that they adapted their Meaſures only to the middle of the refracted Light; ſo that from their Meaſures we may conclude only that the Rays which have a mean degree of Refrangibility , that is thoſe which when ſeparated from the reſt appear green, are refracted according to a given Proportion of their Sines. And therefore we are now to ſhew that the like given Proportions obtain in all the reſt. That it ſhould be ſo is very reaſonable, Nature being ever conformable to her ſelf : but an experimental Proof is deſired. And ſuch a Proof will be had if we can ſhew that the Sines of Refraction of Rays differently Refrangible are one to another in a given Proportion when their Sines of Incidence are equal. For if the Sines of Refraction of all the Rays are in given Proportions to the Sine of Refraction

of

of a Ray which has a mean degree of Refrangibility, and this Sine is in a given Proportion to the equal Sines of Incidence, thofe other Sines of Refraction will alfo be in given Proportions to the equal Sines of Incidence. Now when the Sines of Incidence are equal, it will appear by the following Experiment that the Sines of Refraction are in a given Proportion to one another.

Exper. 15. The Sun fhining into a dark Chamber through a little round hole in the Window-fhut, let S re- *Fig.* 26. prefent his round white Image painted on the oppofite Wall by his direct Light, P T his oblong coloured Image made by refracting that Light with a Prifm placed at the Window; and *pt*, or *2 p 2 t*, or *3 p 3 t*, his oblong coloured Image made by refracting again the fame Light fideways with a fecond Prifm placed immediately after the firft in a crofs Pofition to it, as was explained in the fifth Experiment : that is to fay, *pt* when the Refraction of the fecond Prifm is fmall, *2 p 2 t* when its Refraction is greater, and *3 p 3 t* when it is greateft. For fuch will be the diverfity of the Refractions if the refracting Angle of the fecond Prifm be of various Magnitudes ; fuppofe of fifteen or twenty degrees to make the Image *p t*, of thirty or forty to make the Image *2 p 2 t*, and of fixty to make the Image *3 p 3 t*. But for want of folid Glafs Prifms with Angles of convenient bigneffes, there may be Veffels made of polifhed Plates of Glafs cemented together in the form of Prifms and filled with Water. Thefe things being thus ordered, I obferved that all the folar Images or coloured Spectrums P T, *pt*, *2 p 2 t*, *3 p 3 t* did very nearly converge to the place S on which the direct Light of the Sun fell and painted his white round Image when the Prifms were taken away. The Axis of the Spectrum P T, that is the Line drawn through the middle of it Parallel to

its

its Rectilinear Sides, did when produced pass exactly through the middle of that white round Image S. And when the Refraction of the second Prism was equal to the Refraction of the first, the refracting Angles of them both being about 60 degrees, the Axis of the Spectrum $3p$ $3t$ made by that Refraction, did when produced pass also through the middle of the same white round Image S. But when the Refraction of the second Prism was less than that of the first, the produced Axes of the Spectrums tp or $2t$ $2p$ made by that Refraction did cut the produced Axis of the Spectrum TP in the Points m and n, a little beyond the Center of that white round Image S. Whence the Proportion of the Line $3t$T to the Line $3p$P was a little greater than the Proportion of $2t$T to $2p$P, and this Proportion a little greater than that of tT to pP. Now when the Light of the Spectrum P T falls perpendicularly upon the Wall, those Lines $3t$T, $3p$P, and $2t$T, $2p$P and tT, pP, are the Tangents of the Refractions; and therefore by this Experiment the Proportions of the Tangents of the Refractions are obtained, from whence the Proportions of the Sines being derived, they come out equal, so far as by viewing the Spectrums and using some Mathematical reasoning I could Estimate. For I did not make an Accurate Computation. So then the Proposition holds true in every Ray apart, so far as appears by Experiment. And that it is accurately true may be demonstrated upon this Supposition, *That Bodies refract Light by acting upon its Rays in Lines Perpendicular to their Surfaces.* But in order to this Demonstration, I must distinguish the Motion of every Ray into two Motions, the one Perpendicular to the refracting Surface, the other Parallel to it, and concerning the Perpendicular Motion lay down the following Proposition.

If

If any Motion or moving thing whatfoever be incident with any velocity on any broad and thin Space terminated on both fides by two Parallel Planes, and in its paffage through that fpace be urged perpendicularly towards the further Plane by any force which at given diftances from the Plane is of given quantities; the perpendicular Velocity of that Motion or Thing, at its emerging out of that fpace, fhall be always equal to the Square Root of the Summ of the Square of the perpendicular Velocity of that Motion or Thing at its Incidence on that fpace; and of the Square of the perpendicular Velocity which that Motion or Thing would have at its Emergence, if at its Incidence its perpendicular Velocity was infinitely little.

And the fame Propofition holds true of any Motion or Thing perpendicularly retarded in its paffage through that fpace, if inftead of the Summ of the two Squares you take their difference. The Demonftration Mathematicians will eafily find out, and therefore I fhall not trouble the Reader with it.

Suppofe now that a Ray coming moft obliquely in the *Fig.* 1. Line MC be refracted at C by the Plane RS into the Line CN, and if it be required to find the Line CE into which any other Ray AC fhall be refracted; let MC, AD, be the Sines of incidence of the two Rays, and NG, EF, their Sines of Refraction, and let the equal Motions of the Incident Rays be reprefented by the equal Lines M C and AC, and the Motion MC being confidered as parallel to the refracting Plane, let the other Motion AC be diftinguifhed into two Motions AD and DC, one of which AD is parallel, and the other DC perpendicular to the refracting Surface. In like manner, let the Motions of the emering Rays be diftinguifh'd into two, whereof the perpendicular

H   pendicular

perpendicular ones are $\frac{MC}{NG}$ CG and $\frac{AD}{EF}$ CF. And if the force of the refracting Plane begins to act upon the Rays either in that Plane or at a certain distance from it on the one side, and ends at a certain distance from it on the other side, and in all places between those two Limits acts upon the Rays in Lines perpendicular to that rafracting Plane, and the Actions upon the Rays at equal distances from the refracting Plane be equal, and at unequal ones either equal or unequal according to any rate whatever ; that motion of the Ray which is Parallel to the refracting Plane will suffer no alteration by that force ; and that motion which is perpendicular to it will be altered according to the rule of the foregoing Proposition. If therefore for the perpendicular Velocity of the emerging Ray CN you write $\frac{MC}{NG}$ CG as above, then the perpendicular Velocity of any other emerging Ray CE which was $\frac{AD}{EF}$ CF, will be equal to the square Root of $CDq + \frac{MCq}{NGq}$ CGq. And by squaring these equals, and adding to them the Equals ADq and MCq—CDq, and dividing the Summs by the Equals CFq + EFq and CGq + NGq, you will have $\frac{ADq}{EFq}$ equal to $\frac{MCq}{NGq}$. Whence AD, the Sine of Incidence, is to EF the Sine of Refraction, as MC to NG, that is, in a given *ratio*. And this Demonstration being general, without determining what Light is, or by what kind of force it is refracted, or assuming any thing further than that the refracting Body acts upon the Rays in Lines perpendicular to its Surface ; I take it to be a very convincing Argument of the full Truth of this Proposition.

So

So then, if the *ratio* of the Sines of Incidence and Refraction of any fort of Rays be found in any one Cafe, 'tis given in all Cafes; and this may be readily found by the Method in the following Propofition.

P R O P. VII.   Theor. VI.

The Perfection of Telefcopes is impeded by the different Refrangibility of the Rays of Light.

THE imperfection of Telefcopes is vulgarly attributed to the fpherical Figures of the Glaffes, and therefore Mathematicians have propounded to Figure them by the Conical Sections. To fhew that they are miftaken, I have inferted this Propofition; the truth of which will appear by the meafures of the Refractions of the feveral forts of Rays ; and thefe meafures I thus determine.

In the third experiment of the firft Book, where the refracting Angle of the Prifm was 62¼ degrees, the half of that Angle 31 deg. 15 min. is the Angle of Incidence of the Rays at their going out of the Glafs into the Air ; and the Sine of this Angle is 5188, the Radius being 10000. When the Axis of this Prifm was parallel to the Horizon, and the Refraction of the Rays at their Incidence on this Prifm equal to that at their Emergence out of it, I obferved with a Quadrant the Angle which the mean refrangible Rays (that is, thofe which went to the middle of the Sun's coloured Image) made with the Horizon and by this Angle and the Sun's altitude obferved at the fame time, I found the Angle which the emergent Rays contained with the incident to be 44 deg. and 40 min. and the half of this Angle added to the Angle of Incidence 31 deg. 15 min. makes the

Angle

Angle of Refraction, which is therefore 53 deg. 35 min. and its Sine 8047. Thefe are the Sines of Incidence and Refraction of the mean refrangible Rays, and their proportion in round numbers is 20 to 31. This Glafs was of a colour inclining to green. The laft of the Prifms mentioned in the third Experiment was of clear white Glafs. Its refracting Angle 63½ degrees. The Angle which the emergent Rays contained, with the incident 45 deg. 50 min. The Sine of half the firft Angle 5262. The Sine of half the Summ of the Angles 8157. And their proportion in round numbers 20 to 31 as before.

From the Length of the Image, which was about 9¼ or 10 Inches, fubduct its Breadth, which was 2⅛ Inches, and the Remainder 7¾ Inches would be the length of the Image were the Sun but a point, and therefore fubtends the Angle which the moft and leaft refrangible Rays, when incident on the Prifm in the fame Lines, do contain with one another after their Emergence. Whence this Angle is 2 deg. 0.′ 7.″ For the diftance between the Image and the Prifm where this Angle is made, was 18½ Feet, and at that diftance the Chord 7¾ Inches fubtends an Angle of 2 deg. 0.′ 7.″ Now half this Angle is the Angle which thefe emergent Rays contain with the emergent mean refrangible Rays, and a quarter thereof, that is 30.′ 2.″ may be accounted the Angle which they would contain which the fame emergent mean refrangible Rays, were they co-incident to them within the Glafs and fuffered no other Refraction then that at their Emergence. For if two equal Refractions, the one at the incidence of the Rays on the Prifm, the other at their Emergence, make half the Angle 2 deg. 0.′ 7.″ then one of thofe Refractions will make about a quarter of that Angle, and this quarter added to

and

and subducted from the Angle of Refraction of the mean refrangible Rays, which was 53 deg. 35', gives the Angles of Refraction of the moſt and leaſt refrangible Rays 54 deg. 5' 2", and 53 deg. 4' 58", whoſe Sines are 8099 and 7995, the common Angle of Incidence being 31 deg. 15' and its Sine 5188; and theſe Sines in the leaſt round numbers are in proportion to one another as 78 and 77 to 50.

Now if you ſubduct the common Sine of Incidence 50 from the Sines of Refraction 77 and 78, the remainders 27 and 28 ſhew that in ſmall Refractions the Refraction of the leaſt refrangible Rays is to the Refraction of the moſt refrangible ones as 27 to 28 very nearly, and that the difference of the Refractions of the leaſt refrangible and moſt refrangible Rays is about the $27\frac{1}{2}$th part of the whole Refraction of the mean refrangible Rays.

Whence they that are skilled in Opticks will eaſily underſtand, that the breadth of the leaſt circular ſpace into which Object-Glaſſes of Teleſcopes can collect all ſorts of Parallel Rays, is about the $27\frac{1}{2}$th part of half the aperture of the Glaſs, or 55th part of the whole aperture; and that the Focus of the moſt refrangible Rays is nearer to the Object-Glaſs than the Focus of the leaſt refrangible ones, by about the $27\frac{1}{2}$th part of the diſtance between the Object-Glaſs and the Focus of the mean refrangible ones.

And if Rays of all ſorts, flowing from any one lucid point in the Axis of any convex Lens, be made by the Refraction of the Lens to converge to points not too remote from the Lens, the Focus of the moſt refrangible Rays ſhall be nearer to the Lens than the Focus of the leaſt refrangible ones, by a diſtance which is to the $27\frac{1}{2}$th part of the diſtance of the Focus of the mean refrangible Rays from the Lens as the diſtance between that Focus and the lucid

point

point from whence the Rays flow is to the diftance between that lucid point and the Lens very nearly.

Now to examine whether the difference between the Refractions which the moft refrangible and the leaft refrangible Rays flowing from the fame point fuffer in the Object-Glaffes of Telefcopes and fuch like Glaffes, be fo great
as is here defcribed, I contrived the following Experiment.

Exper. 16. The Lens which I ufed in the fecond and
eighth Experiments, being placed fix Feet and an Inch diftant from any Object, collected the Species of that Object
by the mean refrangible Rays at the diftance of fix Feet
and an Inch from the Lens on the other fide. And therefore by the foregoing Rule it ought to collect the Species of
that Object by the leaft refrangible Rays at the diftance of
fix Feet and $3\frac{2}{3}$ Inches from the Lens, and by the moft refrangible ones at the diftance of five Feet and $10\frac{1}{3}$ Inches
from it : So that between the two Places where thefe leaft
and moft refrangible Rays collect the Species, there may
be the diftance of about $5\frac{1}{3}$ Inches. For by that Rule, as
fix Feet and an Inch (the diftance of the Lens from the
lucid Object) is to twelve Feet and two Inches (the diftance of the lucid Object from the Focus of the mean refrangible Rays) that is, as one is to two, fo is the $27\frac{1}{2}$th
part of fix Feet and an Inch (the diftance between the Lens
and the fame Focus) to the diftance between the Focus of
the moft refrangible Rays and the Focus of the leaft refrangible ones, which is therefore $5\frac{17}{55}$ Inches, that is very
nearly $5\frac{1}{3}$ Inches. Now to know whether this meafure
was true, I repeated the fecond and eighth Experiment of
this Book with coloured Light, which was lefs compounded than that I there made ufe of : For I now feparated the

hetero-

h eterogeneous Rays from one another by the Method I described in the 11th Experiment, so as to make a coloured
Spectrum about twelve or fifteen times longer than broad.
This Spectrum I cast on a printed book, and placing the
above-mentioned Lens at the distance of six Feet and an
Inch from this Spectrum to collect the Species of the illuminated Letters at the same distance on the other side, I
found that the Species of the Letters illuminated with Blue
were nearer to the Lens than those illuminated with deep
Red by about three Inches or three and a quarter: but the
Species of the Letters illuminated with Indigo and Violet
appeared so confused and indistinct, that I could not read
them : Whereupon viewing the Prism, I found it was full
of Veins running from one end of the Glass to the other ;
so that the Refraction could not be regular. I took another Prism therefore which was free from Veins, and instead of the Letters I used two or three Parallel black Lines
a little broader than the stroakes of the Letters, and casting the Colours upon these Lines in such manner that the
Lines ran along the Colours from one end of the Spectum
to the other, I found that the Focus where the Indigo, or
confine of this colour and Violet cast the Species of the
black Lines most distinctly, to be about 4 Inches or 4¼ nearer to the Lens than the Focus where the deepest Red cast
the Species of the same black Lines most distinctly.
The violet was so faint and dark, that I could not
discern the Species of the Lines distinctly by that Colour ; and therefore considering that the Prism was made
of a dark coloured Glass inclining to Green, I took another
Pism of clear white Glass ; but the Spectrum of Colours
which this Prism made had long white Streams of faint
Light shooting out from both ends of the Colours, which
made me conclude that something was amiss ; and viewing

ing the Prifm, I found two or three little Bubbles in the
Glafs which refracted the Light irregularly. Wherefore I
covered that part of the Glafs with black Paper, and letting the Light pafs through another part of it which was
free from fuch Bubles, the Spectrum of Colours became
free from thofe irregular Streams of Light, and was now
fuch as I defired. But ftill I found the Violet fo dark and
faint, that I could fcarce fee the Species of the Lines by the
Violet, and not at all by the deepeft part of it, which was
next the end of the Spectrum. I fufpected therefore that
this faint and dark Colour might be allayed by that fcattering Light which was refracted, and reflected irregularly
partly by fome very fmall Bubbles in the Glaffes and
partly by the inequalities of their Polifh: which Light,
tho' it was but little, yet it being of a White Colour,
might fuffice to affect the Senfe fo ftrongly as to difturb
the Phænomena of that weak and dark Colour the Violet,
and therefore I tried, as in the 12th, 13th, 14th Experiments, whether the Light of this Colour did not confift of
a fenfible mixture of heterogeneous Rays, but found it did
not. Nor did the Refractions caufe any other fenfible
Colour than Violet to emerge out of this Light, as they
would have done out of White Light, and by confequence out of this Violet Light had it been fenfibly compounded with White Light. And therefore I concluded, that the reafon why I could not fee the Species of
the Lines diftinctly by this Colour, was only the darknefs
of this Colour and Thinnefs of its Light, and its diftance from the Axis of the Lens; I divided therefore thofe
Parallel Black Lines into equal Parts, by which I might
readily know the diftances of the Colours in the Spectrum
from one another, and noted the diftances of the Lens
from the Foci of fuch Colours as caft the Species of the

Lines

Lines diftinctly, and then confidered whether the difference of thofe diftances bear fuch proportion to 5¼ Inches,
the greateft difference of the diftances which the Foci of
the deepeft Red and Violet ought to have from the Lens,
as the diftance of the obferved Colours from one another
in the Spectrum bear to the like diftance of the deepeft Red
and Violet meafured in the rectilinear fides of the Spectrum, that is, to the length of thofe fides or excefs of the
length of the Spectrum above its breadth. And my Obfervations were as follows.

When I obferved and compared the deepeft fenfible Red,
and the Colour in the confine of Green and Blue, which
at that rectilinear fides of the Spectrum was diftant from it
half the length of thofe fides, the Focus where the confine
of Green and Blue caft the Species of the Lines diftinctly
on the Paper, was nearer to the Lens then the Focus where
the Red caft thofe Lines diftinctly on it by about $2\frac{1}{2}$ or
$2\frac{3}{4}$ Inches. For fometimes the Meafures were a little greater, fometimes a little lefs, but feldom varied from one
another above $\frac{1}{3}$ of an Inch. For it was very difficult to
define the Places of the Foci, without fome little Errors.
Now if the Colours diftant half the length of the Image,
(meafured at its rectilinear fides) give $2\frac{1}{2}$ or $2\frac{3}{4}$ difference
of the diftances of their Foci from the Lens, then the Co-
lours diftant the whole length ought to give 5 or $5\frac{1}{2}$ Inches
difference of thofe diftances.

But here it's to be noted, that I could not fee the Red
to the full End of the Spectrum, but only to the Center
of the Semicircle which bounded that End, or a little farther ; and therefore I compared this Red not with that Colour which was exactly in the middle of the Spectrum, or
confine of Green and Blue, but with that which verged a
little more to the Blue than to the Green : And as I reck-

I   oned

oned the whole length of the Colours not to be the whole length of the Spectrum, but the length of its rectilinear sides, so completing the Semicirlar Ends into Circles, when either of the observed Colours fell within those Circles, I measured the distance of that Colour from the End of the Spectrum, and subducting half the distance from the measured distance of the Colours, I took the remainder for their corrected distance ; and in these Observations set down this corrected distance for the difference of their distances from the Lens. For as the length of the rectilinear sides of the Spectrum would be the whole length of all the Colours, were the Circles of which (as we shewed) that Spectrum consists contracted and reduced to Physical Points, so in that Case this corrected distance would be the real distance of the observed Colours.

When therefore I further observed the deepest sensible Red, and that Blue whose corrected distance from it was $\frac{7}{12}$ parts of the length of the rectilinear sides of the Spectrum, the difference of the distances of their Foci from the Lens was about $3\frac{1}{4}$ Inches, and as 7 to 12 so is $3\frac{1}{4}$ to $5\frac{4}{7}$.

When I observed the deepest sensible Red, and that Indigo whose corrected distance was $\frac{8}{12}$ or $\frac{2}{3}$ of the length of the rectilinear sides of the Spectrum, the difference of the distances of their Foci from the Lens, was about $3\frac{2}{3}$ Inches, and as 2 to 3 so is $3\frac{2}{3}$ to $5\frac{1}{2}$.

When I observed the deepest sensible Red, and that deep Indigo whose corrected distance from one another was $\frac{9}{12}$ or $\frac{3}{4}$ of the length of the rectilinear sides of the Spectum, the difference of the distances of their Foci from the Lens was about 4 Inches ; and as 3 to 4 so is 4 to $5\frac{1}{3}$.

When I observed the deepest sensible Red, and that part of the Violet next the Indigo whose corrected distance from the Red was $\frac{10}{12}$ or $\frac{5}{6}$ of the length of the rectilinear sides of

the

the Spectrum, the difference of the diſtances of their Foci from the Lens was about $4\frac{1}{2}$ Inches; and as 5 to 6, ſo is $4\frac{1}{2}$ to $5\frac{2}{5}$. For ſometimes when the Lens was advantagiouſly placed, ſo that its Axis reſpected the Blue, and all things elſe were well ordered, and the Sun ſhone clear, and I held my Eye very near to the Paper on which the Lens caſt the Species of the Lines, I could ſee pretty diſtinctly the Species of thoſe Lines by that part of the Violet which was next the Indigo ; and ſometimes I could ſee them by above half the Violet. For in making theſe Experiments I had obſerved, that the Species of thoſe Colours only appeared diſtinct which were in or near the Axis of the Lens : So that if the Blue or Indigo were in the Axis, I could ſee their Species diſtinctly ; and then the Red appeared much leſs diſtinct than before. Wherefore I contrived to make the Spectrum of Colours ſhorter than before, ſo that both its Ends might be nearer to the Axis of the Lens. And now its length was about $2\frac{1}{2}$ Inches and breadth about $\frac{1}{5}$ or $\frac{1}{6}$ of an Inch. Alſo inſtead of the black Lines on which the Spectrum was caſt, I made one black Line broader than thoſe, that I might ſee its Species more eaſily ; and this Line I divided by ſhort croſs Lines into equal Parts, for meaſuring the diſtances of the obſerved Colours. And now I could ſometimes ſee the Species of this Line with its diviſions almoſt as far as the Centers of the Semicircular Violet End of the Spectrum, and made theſe further Obſervations.

When I obſerved the deepeſt ſenſible Red, and that part of the Violet whoſe corrected diſtance from it was about $\frac{8}{9}$ Parts of the rectilinear ſides of the Spectrum the difference of the diſtances of the Foci of thoſe Colours from the Lens, was one time $4\frac{2}{3}$, another time $4\frac{3}{4}$, another time $4\frac{7}{8}$, Inches, and as 8 to 9, ſo are $4\frac{2}{3}$, $4\frac{3}{4}$, $4\frac{7}{8}$, to $5\frac{1}{4}$, $5\frac{11}{32}$, $5\frac{31}{64}$ reſpectively.

I 2

When

When I obferved the deepeft fenfible Red, and deepeft
fenfible Violet, (the corrected diftance of which Colours
when all things were ordered to the beft advantage, and the
Sun fhone very clear, was about $\frac{11}{12}$ or $\frac{15}{16}$ parts of the length
of the rectilinear fides of the coloured Spectrum,) I found
the difference of the diftances of their Foci from the Lens
fometimes $4\frac{3}{4}$ fometimes $5\frac{1}{4}$, and for the moft part 5 Inches
or thereabouts : and as 11 to 12 or 15 to 26, fo is five
Inches to $5\frac{1}{2}$ or $5\frac{2}{3}$ Inches.

And by this progreffion of Experiments I fatisfied my
felf, that had the light at the very Ends of the Spectrum been
ftrong enough to make the Species of the black Lines appear plainly on the Paper, the Focus of the deepeft Violet would have been found nearer to the Lens, than the Focus of the deepeft Red, by about $5\frac{1}{3}$ Inches at leaft. And
this is a further Evidence, that the Sines of Incidence and
Refraction of the feveral forts of Rays, hold the fame proportion to one another in the fmalleft Refractions which
they do in the greateft.

My progrefs in making this nice and troublefome Experiment I have fet down more at large, that they that fhall
try it after me may be aware of the Circumfpection requifite to make it fucceed well. And if they cannot make
it fucceed fo well as I did, they may notwithftanding collect by the Proportion of the diftance of the Colours in the
Spectrum, to the difference of the diftances of their Foci
from the Lens, what would be the fuccefs in the more diftant Colours by a better Trial. And yet if they ufe a
broader Lens than I did, and fix it to a long ftreight Staff
by means of which it may be readily and truly directed to
the Colour whofe Focus is defired, I queftion not but the
Experiment will fucceed better with them than it did with
me. For I directed the Axis as nearly as I could to the
middle

middle of the Colours, and then the faint Ends of the Spectrum being remote from the Axis, cast their Species less distinctly on the Paper than they would have done had the Axis been successively directed to them.

Now by what has been said its certain, that the Rays which differ in refrangibility do not converge to the same Focus, but if they flow from a lucid point, as far from the Lens on one side as their Foci are one the other, the Focus of the most refrangible Rays shall be nearer to the Lens than that of the least refrangible, by above the four-teenth part of the whole distance: and if they flow from a lu-cid point, so very remote from the Lens that before their Incidence they may be accounted Parallel, the Focus of the most refrangible Rays shall be nearer to the Lens than the Focus of the least refrangible, by about the 27th or 28th part of their whole distance from it. And the Diameter of the Circle in the middle space between those two Foci which they illuminate when they fall there on any Plane, perpen-dicular to the Axis (which Circle is the least into which they can all be gathered) is about the 55th part of the Dia-meter of the aperture of the Glass. So that 'tis a wonder that Telescopes represent Objects so distinct as they do. But were all the Rays of Light equally refrangible, the Error arising only from the sphericalness of the Figures of Glasses would be many hundred times less. For if the Object-Glass of a Telescope be Plano-convex, and the Plane side be turned towards the Object, and the Diameter of the Sphere whereof this Glass is a segment, be called D, and the Semidiameter of the aperture of the Glass be called S, and the Sine of Incidence out of Glass into Air, be to the Sine of Refraction as I to R: the Rays which come Parallel to the Axis of the Glass, shall in the Place where the Image of the Object is most distinctly made, be scattered all over a little

Circle

Circle whose Diameter is $\frac{R}{I} \times \frac{S \, cub.}{D \, quad.}$ very nearly, as I gather by computing the Errors of the Rays by the method
of infinite Series, and rejecting the Terms whose quanti-
tities are inconsiderable. As for instance, if the Sine of Incidence I, be to the Sine of Refraction R, as 20 to 31, and
if D the Diameter of the Sphere to which the Convex side
of the Glass is ground, be 100 Feet or 1200 Inches, and
S the Semidiameter of the aperture be two Inches, the
Diameter of the little Circle (that is $\frac{R \times S \, cub.}{I \times D \, quad.}$) will be
$\frac{31 \times 8}{20 \times 1200 \times 1200}$ (or $\frac{31}{3600000}$) parts of an Inch. But the
Diameter of the little Circle through which these Rays are
scattered by unequal refrangibility, will be about the 55th
part of the aperture of the Object-Glass which here is four
Inches. And therefore the Error arising from the spherical
Figure of the Glass, is to the Error arising from the diffe-
rent Refrangibility of the Rays, as $\frac{31}{3600000}$ to $\frac{4}{55}$ that is as 1
to 8151 : and therefore being in Comparison so very little,
deserves not to be considered.

But you will say, if the Errors caused by the different refrangibility be so very great, how comes it to pass that Objects appear through Telescopes so distinct as they do ? I answer, 'tis because the erring Rays are not scattered uniformly over all that circular space, but collected infinitely more
densely in the Center than in any other part of the Circle,
and in the way from the Center to the Circumference grow
continually rarer and rarer, so as at the Circumference to
become infinitely rare; and by reason of their rarity are
not strong enough to be visible, unless in the Center and very near it. Let ADE represent one of those Circles described with the Center C and Semidiameter AC, and let
BFG be a smaller Circle concentric to the former, cutting
with

Fig. 27.

with its Circumference the Diameter AC in B, and befect AC in N, and by my reckoning the denfity of the Light in any place B will be to its denfity in N, as AB to BC; and the whole Light within the leffer Circle BFG, will be to the whole Light within the greater AED, as the Excefs of the Square of AC above the Square of AB, is to the Square of AC. As if BC be the fifth part of AC, the Light will be four times denfer in B than in N, and the whole Light within the lefs Circle, will be to the whole Light within the greater, as nine to twenty five. Whence it's evident that the Light within the lefs Circle, muft ftrike the fenfe much more ftrongly, than that faint and dilated light round about between it and the Circumference of the greater.

But its further to be noted, that the moft luminous of the prifmatick Colours are the Yellow and Orange. Thefe affect the Senfes more ftrongly than all the reft together, and next to thefe in ftrength are the Red and Green. The Blue compared with thefe is a faint and dark Colour, and the Indigo and Violet are much darker and fainter, fo that thefe compared with the ftronger Colours are little to be regarded. The Images of Objects are therefore to be placed, not in the Focus of the mean refrangible Rays which are in the confine of Green and Blue, but in the Focus of thofe Rays which are in the middle of the Orange and Yellow; there where the Colour is moft luminous and fulgent, that is in the brighteft Yellow, that Yellow which inclines more to Orange than to Green. And by the Refraction of thefe Rays (whofe Sines of Incidence and Refraction in Glafs are as 17 and 11) the Refraction of Glafs and Cryftal for optical ufes is to be meafured. Let us therefore place the Image of the Object in the Focus of thefe Rays, and all the Yellow and Orange will fall within a Circle, whofe Diameter is about the 25oth part of the Diameter of the aperture

ture of the Glafs. And if you add the brighter half of the
Red, (that half which is next the Orange, and the brighter
half of the Green, (that half which is next the Yellow,) a-
bout three fifth parts of the Light of thefe two Colours will
fall within the fame Circle, and two fifth parts will fall without it round about; and that which falls without will be
fpread through almoft as much more fpace as that which
falls within, and fo in the grofs be almoft three times rarer. Of the other half of the Red and Green, (that is of
the deep dark Red and Willow Green) about one quarter
will fall within this Circle, and three quarters without, and
that which falls without will be fpread through about four
or five times more fpace than that which fall within; and fo
in the grofs be rarer, and if compared with the whole Light
within it, will be about 25 times rarer than all that taken in
the grofs ; or rather more than 30 or 40 times rarer, becaufe the deep red in the end of the Spectrum of Colours
made by a Prifm is very thin and rare, and the Willow Green
is fomething rarer than the Orange and Yellow. The Light
of thefe Colours therefore bring fo very much rarer than that
within the Circle, will fcarce affect the Senfe efpecially fince
the deep Red and Willow Green of this Light, are much
darker Colours then the reft. And for the fame reafon the
Blue and Violet being much darker Colours than thefe, and
much more rarified, may be neglected. For the denfe and
bright Light of the Circle, will obfcure the rare and weak
Light of thefe dark Colours round about it, and render them
almoft infenfible. The fenfible Image of a lucid point is
therefore fcarce broader than a Circle whofe Diameter is
the 250th part of the diameter of the aperture of the Object
Glafs of a good Telefcope, or not much broader, if you
except a faint and dark mifty light round about it, which
a Spectator will fcarce regard. And therefore in a Telefcope
whofe

whofe aperture is four Inches, and length an hundred Feet, it exceeds not $2'' 45'''$, or $3''$. And in a Telefcope whofe aperture is two Inches, and length 20 or 30 Feet, it may be $5''$ or $6''$ and fcarce above. And this Anfwers well to Experience : For fome Aftronomers have found the Diameters of the fixt Stars, in Telefcopes of between twenty and fixty Feet in length, to be about $4''$ or $5''$ or at moft $6''$ in Diameter. But if the Eye-Glafs be tincted faintly with the fmoke of a Lamp or Torch, to obfcure the Light of the Star, the fainter Light in the circumference of the Star ceafes to be vifible, and the Star (if the Glafs be fufficiently foiled with fmoke) appears fomething more like a Mathematical Point. And for the fame reafon, the enormous part of the Light in the Circumference of every lucid Point ought to be lefs difcernable in fhorter Telefcopes than in longer, becaufe the fhorter tranfmit lefs Light to the Eye.

Now if we fuppofe the fenfible Image of a lucid point, to be even 250 times narrower than the aperture of the Glafs: yet were it not for the different refrangibility of the Rays, its breadth in an 100 Foot Telefcope whofe aperture is 4 Inches would be but $\frac{31}{3600000}$ parts of an Inch, as is manifeft by the foregoing Computation. And therefore in this Cafe the greateft Errors arifing from the fpherical Figure of the Glafs, would be to the greateft fenfible Errors arifing from the different refrangibility of the Rays as $\frac{11}{3600000}$ to $\frac{4}{250}$ at moft, that is only as 1 to 1826. And this fufficiently fhews that it is not the fpherical Figures of Glaffes but the different refrangibility of the Rays which hinders the perfection of Telefcopes.

There is another Argument by which it may appear that the different refrangibility of Rays, is the true Caufe of the imperfection of Telefcopes. For the Errors of the Rays arifing from the fpherical Figures of Object-Glaffes, are as

K

the

the Cubes of the apertures of the Object-Glasses; and thence to make Telescopes of various lengths, magnify with equal distinctness, the apertures of the Object-Glasses, and the Charges or magnifying Powers, ought to be as the Cubes of the square Roots of their lengths; which doth not answer to Experience. But the errors of the Rays arising from the different refrangibility, are as the apertures of the Object-Glasses, and thence to make Telescopes of various lengths, magnify with equal distinctness, their apertures and charges ought to be as the square Roots of their lengths; and this answers to experience as is well known. For instance, a Telescope of 64 Feet in length, with an aperture of $2\frac{2}{3}$ Inches, magnifies about 120 times, with as much distinctness as one of a Foot in length, with $\frac{1}{3}$ of an Inch aperture, magnifies 15 times.

Now were it not for this different refrangibility of Rays, Telescopes might be brought to a greater Perfection than we have yet described, by composing the Object-Glass of two Glasses with Water between them. *Fig. 28.* Let ADFC represent the Object-Glass composed of two Glasses ABED and and BEFC, alike convex on the outsides AGD and CHF, and alike concave on the insides BME, BNE, with Water in the concavity BMEN. Let the Sine of Incidence out of Glass into Air be as I to R and out of Water into Air as K to R, and by consequence out of Glass into Water, as I to K : and let the Diameter of the Sphere to which the convex sides AGD and CHF are ground be D, and the Diameter of the Sphere to which the concave sides BME and BNE are ground be to D, as the Cube Root of KK—KI to the Cube Root of RK — RI: and the Refractions on the concave sides of the Glasses, will very much correct the Errors of the Refractions on the convex sides, so far as they arise from the sphericalness of the Figure. And by this means might

might Telescopes be brought to sufficient perfection, were it not for the different refrangibility of several sorts of Rays. But by reason of this different refrangibility, I do not yet see any other means of improving Telescopes by Refractions alone than that of increasing their lengths, for which end the late contrivance of *Hugenius* seems well accommodated. For very long Tubes are cumbersome, and scarce to be readily managed, and by reason of their length are very apt to bend, and shake by bending so as to cause a continual trembling in the Objects, whereby it becomes difficult to see them distinctly : whereas by his contrivance the Glasses are readily manageable, and the Object-Glass being fixt upon a strong upright Pole becomes more steddy.

Seeing therefore the improvement of Telescopes of given lengths by Refractions is desperate ; I contrived heretofore a Perspective by reflexion, using instead of an Object Glass a concave Metal. The diameter of the Sphere to which the Metal was ground concave was about 25 English Inches, and by consequence the length of the Instrument about six Inches and a quarter. The Eye-Glass was plano-convex, and the Diameter of the Sphere to which the convex side was ground was about ⅕ of an Inch, or a little less, and by consequence it magnified between 30 and 40 times. By another way of measuring I found that it magnified about 35 times. The Concave Metal bore an aperture of an Inch and a third part ; but the aperture was limited not by an opake Circle, covering the Limb of the Metal round about, but by an opake circle placed between the Eye-Glass and the Eye, and perforated in the middle with a little round hole for the Rays to pass through to the Eye. For this Circle by being placed here, stopt much of the erroneous Light, which otherwise would have disturbed the Vision. By comparing it with a pretty good Perspective of four Feet in

K 2

length, made with a concave Eye-Glaſs, I could read at a greater diſtance with my own Inſtrument than with the Glaſs. Yet Objects appeared much darker in it than in the Glaſs, and that partly becauſe more Light was loſt by reflexion in the Metal, then by refraction in the Glaſs, and partly becauſe my Inſtrument was overcharged. Had it magnified but 30 or 25 times it would have made the Object appear more briſk and pleaſant. Two of theſe I made about 16 Years ago, and have one of them ſtill by me by which I can prove the truth of what I write. Yet it is not ſo good as at the firſt. For the concave has been divers times tarniſhed and cleared again, by rubbing it with very ſoft Leather. When I made theſe, an Artiſt in *London* undertook to imitate it ; but uſing another way of poliſhing them than I did, he fell much ſhort of what I had attained to, as I afterwards underſtood by diſcourſing the under-Workman he had imployed. The Poliſh I uſed was on this manner. I had two round Copper Plates each ſix Inches in Diameter, the one convex the other concave, ground very true to one another. On the convex I ground the Object-Metal or concave which was to be poliſh'd, till it had taken the Figure of the convex and was ready for a Poliſh. Then I pitched over the convex very thinly, by dropping melted pitch upon it and warming it to keep the pitch ſoft, whilſt I ground it with the concave Copper wetted to make it ſpread evenly all over the convex. Thus by working it well I made it as thin as a Groat, and after the convex was cold I ground it again to give it as true a Figure as I could. Then I took Putty which I had made very fine by waſhing it from all its groſſer Particles, and laying a little of this upon the pitch, I ground it upon the Pitch with the concave Copper till it had done making a noiſe ; and then upon the Pitch I ground the Object-Metal with a briſk

Motion

Motion, for about two or three Minutes of time, leaning hard upon it. Then I put fresh Putty upon the Pitch and ground it again till it had done making a noise, and afterwards ground the Object Metal upon it as before. And this Work I repeated till the Metal was polished, grinding it the last time with all my strength for a good while together, and frequently breathing upon the Pitch to keep it moist without laying on any more fresh Putty. The Object-Metal was two Inches broad and about one third part of an Inch thick, to keep it from bending. I had two of these Metals, and when I had polished them both I tried which was best, and ground the other again to see if I could make it better than that which I kept. And thus by many Trials I learnt the way of polishing, till I made those two reflecting Perspectives I spake of above. For this Art of polishing will be better learnt by repeated Practice than by my description. Before I ground the Object Metal on the Pitch, I always ground the Putty on it with the concave Copper till it had done making a noise, because if the Particles of the Putty were not by this means made to stick fast in the Pitch, they would by rolling up and down grate and fret the Object Metal and fill it full of little holes.

But because Metal is more difficult to polish than Glass and is afterwards very apt to be spoiled by tarnishing, and reflects not so much Light as Glass quick-silvered over does: I would propound to use instead of the Metal, a Glass ground concave on the foreside, and as much convex on the backside, and quick-silvered over on the convex side. The Glass must be every where of the same thickness exactly. Otherwise it will make Objects look coloured and indistinct. By such a Glass I tried about five or six Years ago to make a reflecting Telescope of four Feet in length to magnify about 150 times, and I satisfied my self that there wants nothing
thing

thing but a good Artift to bring the defign to Perfection.
For the Glafs being wrought by one of our *London* Artifts
after fuch a manner as they grind Glaffes for Telefcopes,
tho it feemed as well wrought as the Object Glaffes ufe to
be, yet when it was quick-filvered, the reflexion difcovered
innumerable Inequalities all over the Glafs. And by reafon
of thefe Inequalities, Objects appeared indiftinct in this Inftrument. For the Errors of reflected Rays caufed by any
Inequality of the Glafs, are about fix times greater than the
Errors of refracted Rays caufed by the like Inequalities. Yet
by this Experiment I fatisfied my felf that the reflexion on
the concave fide of the Glafs, which I feared would difturb
the vifion, did no fenfible prejudice to it, and by confequence
that nothing is wanting to perfect thefe Telefcopes, but
good Workmen who can grind and polifh Glaffes truly fpherical. An Object-Glafs of a fourteen Foot Telefcope, made
by one of our *London* Artificers, I once mended confiderably, by grinding it on Pitch with Putty, and leaning very eafily on it in the grinding, left the Putty fhould fcratch
it. Whether this way may not do well enough for polifhing thefe reflecting Glaffes, I have not yet tried. But he
that fhall try either this or any other way of polifhing which
he may think better, may do well to make his Glaffes ready for polifhing by grinding them without that violence,
wherewith our *London* Workmen prefs their Glaffes in grinding. For by fuch violent preffure, Glaffes are apt to bend
a little in the grinding, and fuch bending will certainly fpoil
their Figure. To recommend therefore the confideration
of thefe reflecting Glaffes, to fuch Artifts as are curious in
figuring Glaffes, I fhall defcribe this Optical Inftrument in
the following Propofition.

PROP.

PROP. VII. Prob. II.

To shorten Telescopes.

LET ABDC represent a Glass spherically concave on the foreside AB, and as much convex on the backside CD, so that it be every where of an equal thickness. Let it not be thicker on one side than on the other, left it make Objects appear coloured and indistinct, and let it be very truly wrought and quick-silvered over on the backside ; and set in the Tube VXYZ which must be very black within. Let EFG represent a Prism of Glass or Crystal placed near the other end of the Tube, in the middle of it, by means of a handle of Brass or Iron FGK, to the end of which made flat it is cemented. Let this Prism be rectangular at E, and let the other two Angles at F and G be accurately equal to each other, and by consequence equal to half right ones, and let the plane sides FE and GE be square, and by consequence the third side FG a rectangular parallelogram, whose length is to its breath in a subduplicate proportion of two to one. Let it be so placed in the Tube, that the Axis of the Speculum may pass through the middle of the square side EF perpendicularly, and by consequence through the middle of the side F G at an Angle of 45 degrees, and let the side EF be turned towards the Speculum, and the distance of this Prism from the Speculum be such that the Rays of the light PQ, RS, &c. which are incident upon the Speculum in Lines Parallel to the Axis thereof, may enter the Prism at the side EF, and be reflected by the side F G, and thence go out of it through the side GE, to the point T which must be the common Focus of the Speculnm ABDC, and of a Plano-convex Eye-Glass H, through which those Rays must pass to the Eye. And let the Rays at their coming

out

Fig. 29.

out of the Glafs pafs through a fmall round hole, or aperture made in a little Plate of Lead, Brafs, or Silver, wherewith the Glafs is to be covered, which hole muft be no
bigger than is neceffary for light enough to pafs through.
For fo it will render the Object diftinct, the Plate in which
'tis made intercepting all the erroneous part of the Light
which comes from the Verges of the Speculum AB. Such
an Inftrument well made if it be 6 Foot long, (reckoning
the length from the Speculum to the Prifm, and thence to
the Focus T) will bear an aperture of 6 Inches at the Speculum, and magnify between two and three hundred times.
But the hole H here limits the aperture with more advantage, then if the aperture was placed at the Speculum. If
the Inftrument be made longer or fhorter, the aperture muft
be in proportion as the Cube of the fquare Root of the
length, and the magnifying as the aperture. But its convenient that the Speculum be an Inch or two broader than
the aperture at the leaft, and that the Glafs of the Speculum
be thick, that it bend not in the working. The Prifm EFG
muft be no bigger than is neceffary, and its back fide FG
muft not be quick-filvered over. For without quick-filver
it will reflect all the Light incident on it from the Speculum.

In this Inftrument the Object will be inverted, but may
be erected by making the fquare fides EF and EG of the
Prifm EFG not plane but fpherically convex, that the Rays
may crofs as well before they come at it as afterwards between it and the Eye-Glafs. If it be defired that the Inftrument bear a larger aperture, that may be alfo done by com-
pofing the Speculum of two Glaffes with Water between
them.

T H E

Fig. 23.

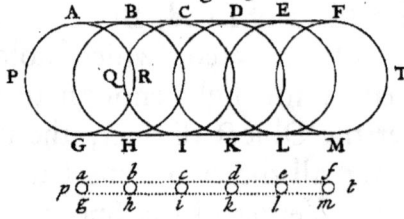

Fig. 24.

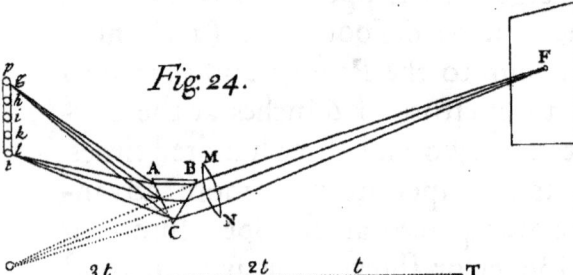

Fig. 25.

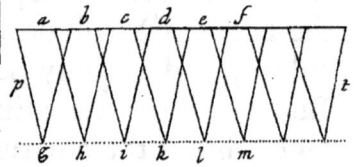

Fig. 26.

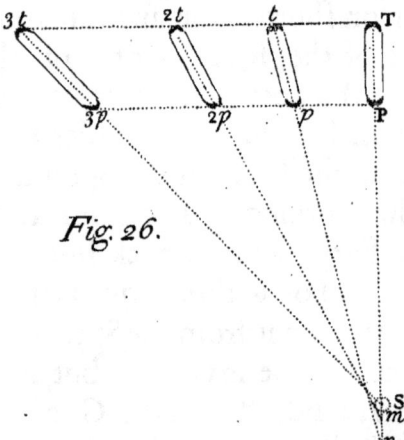

Fig. 27.

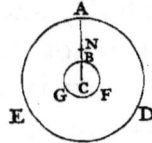

Fig. 28.

Fig. 29.

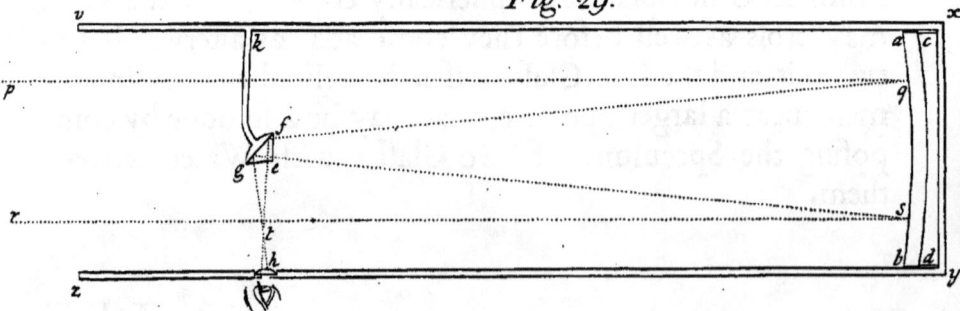

THE

FIRST BOOK

OF

OPTICKS.

PART II.

PROP. I. THEOR. I.

The Phænomena of Colours in refracted or reflected Light
are not caused by new modifications of the Light various-
ly imprest, according to the various terminations of the
Light and Shadow.

The Proof by Experiments.

EXPER. I.

FOR if the Sun shine into a very dark Chamber *Fig.* 1.
through an oblong Hole F, whose breadth is the
sixth or eighth part of an Inch, or something less ; and
his Beam F H do afterwards pass first through a very
large Prism A B C, distant about 20 Feet from the

L

Hole,

Hole, and parallel to it, and then (with its white part) through an oblong Hole H, whose breadth is about the fortieth or sixtieth part of an Inch, and which is made in a black opake Body G I, and placed at the distance of two or three Feet from the Prism, in a parallel situation both to the Prism and to the former Hole, and if this white Light thus transmitted through the Hole H, fall afterwards upon a white Paper p t, placed after that Hole H, at the distance of three or four Feet from it, and there paint the usual Colours of the Prism, suppose red at t, yellow at s, green at r, blue at q, and violet at p ; you may with an iron Wire, or any such like slender opake Body, whose breadth is about the tenth part of an Inch, by intercepting the rays at k, l, m, n or o, take away any one of the Colours at t, s, r, q or p, whilst the other Colours remain upon the Paper as before ; or with an obstacle something bigger you may take away any two, or three, or four Colours together, the rest remaining: So that any one of the Colours as well as violet may become outmost in the confine of the shadow towards p, and any one of them as well as red may become outmost in the confine of the shadow towards t, and any one of them may also border upon the shadow made within the Colours by the obstacle R intercepting some intermediate part of the Light ; and, lastly, any one of them by being left alone may border upon the shadow on either hand. All the Colours have themselves indifferently to any confines of shadow, and therefore the differences of these Colours from one another, do not arise from the different confines of shadow, whereby Light is variously modified as has hitherto been the Opinion of Philosophers.

phers. In trying thefe things 'tis to be obferved, that by how much the Holes F and H are narrower, and the intervals between them, and the Prifm greater, and the Chamber darker, by fo much the better doth the Experiment fucceed ; provided the Light be not fo far diminifhed, but that the Colours at p t be fufficiently vifible. To procure a Prifm of folid Glafs large enough for this Experiment will be difficult, and therefore a prifmatick Veffel muft be made of polifhed Glafs-plates cemented together, and filled with Water.

EXPER. II.

The Sun's Light let into a dark Chamber through *Fig.* 2. the round Hole F, half an Inch wide, paffed firft through the Prifm A B C placed at the Hole, and then through a Lens P T fomething more than four Inches broad, and about eight Feet diftant from the Prifm, and thence converged to O the Focus of the Lens diftant from it about three Feet, and there fell upon a white Paper D E. If that Paper was perpendicular to that Light incident upon it, as 'tis reprefented in the pofture D E, all the Colours upon it at O appeared white. But if the Paper being turned about an Axis parallel to the Prifm, became very much inclined to the Light as 'tis reprefented in the pofitions *de* and *ठ*; the fame Light in the one cafe appeared yellow and red, in the other blue. Here one and the fame part of the Light in one and the fame place, according to the various inclinations of the Paper, appeared in one cafe white, in another yellow or red, in a third blue, whilft the confine of Light and

Shadow,

Shadow, and the refractions of the Prifin in all thefe cafes remained the fame.

EXPER. III.

Fig. 3. Such another Experiment may be more eafily tried as follows. Let a broad beam of the Sun's Light coming into a dark Chamber through a Hole in the Window fhut be refracted by a large Prifm A B C, whofe refracting Angle C is more than 60 degrees, and fo foon as it comes out of the Prifin let it fall upon the white Paper D E glewed upon a ftiff plane, and this Light, when the Paper is perpendicular to it, as 'tis reprefented in D E, will appear perfectly white upon the Paper, but when the Paper is very much inclined to it in fuch a manner as to keep always parallel to the Axis of the Prifm, the whitenefs of the whole Light upon the Paper will according to the inclination of the Paper this way, or that way, change either into yellow and red, as in the pofture *de*, or into blue and violet, as in the pofture *ᵭ͛*. And if the Light before it fall upon the Paper be twice refracted the fame way by two parallel Prifms, thefe Colours will become the more confpicuous. Here all the middle parts of the broad beam of white Light which fell upon the Paper, did without any confine of fhadow to modify it, become coloured all over with one uniform Colour, the Colour being always the fame in the middle of the Paper as at the edges, and this Colour changed according the various obliquity of the reflecting Paper, without any change in the refractions or fhadow, or in the Light which fell upon the Paper. And therefore thefe Colours are

to

to be derived from fome other caufe than the new modifications of Light by refractions and fhadows.

If it be asked, What then is their caufe? I anfwer, That the Paper in the pofture *de*, being more oblique to the more refrangible rays than to the lefs refrangible ones, is more ftrongly illuminated by the latter than by the former, and therefore the lefs refrangible rays are predominant in the reflected Light. And wherever they are predominant in any Light they tinge it with red or yellow, as may in fome meafure appear by the firft Propofition of the firft Book, and will more fully appear hereafter. And the contrary happens in the pofture of the Paper *de*, the more refrangible rays being then predominant which always tinge Light with blues and violets.

EXPER. IV.

The Colours of Bubbles with which Children play are various, and change their fituation varioufly, without any refpect to any confine of fhadow. If fuch a Bubble be covered with a concave Glafs, to keep it from being agitated by any wind or motion of the Air, the Colours will flowly and regularly change their fituation, even whilft the Eye, and the Bubble, and all Bodies which emit any Light, or caft any fhadow, remain unmoved. And therefore their Colours arife from fome regular caufe which depends not on any confine of fhadow. What this caufe is will be fhewed in the next Book.

To

To thefe Experiments may be added the tenth Experiment of the firft Book, where the Sun's Light in a dark Room being trajected through the parallel fuperficies of two Prifms tied together in the form of a Parallelopide, became totally of one uniform yellow or red Colour, at its emerging out of the Prifms. Here, in the production of thefe Colours, the confine of fhadow can have nothing to do. For the Light changes from white to yellow, orange and red fucceffively, without any alteration of the confine of fhadow: And at both edges of the emerging Light where the contrary confines of fhadow ought to produce different effects, the Colour is one and the fame, whether it be white, yellow, orange or red : And in the middle of the emerging Light, where there is no confine of fhadow at all, the Colour is the very fame as at the edges, the whole Light at its very firft emergence being of one uniform Colour, whether white, yellow, orange or red, and going on thence perpetually without any change of Colour, fuch as the confine of fhadow is vulgarly fuppofed to work in refracted Light after its emergence. Neither can thefe Colours arife from any new modifications of the Light by refractions, becaufe they change fucceffively from white to yellow, orange and red, while the refractions remain the fame, and alfo becaufe the refractions are made contrary ways by parallel fuperficies which deftroy one anothers effects. They arife not therefore from any modifications of Light made by refractions and fhadows, but have fome other caufe. What that caufe is we fhewed above in this tenth Experiment, and need not here repeat it.

There

There is yet another material circumftance of this Experiment. For this emerging Light being by a third *Fig.* 22. Prifm H I K refracted towards the Paper P T, and there *Part* 1. painting the ufual Colours of the Prifm, red, yellow, green, blue, violet : If thefe Colours arofe from the refractions of that Prifm modifying the Light, they would not be in the Light before its incidence on that Prifm. And yet in that Experiment we found that when by turning the two firft Prifms about their common Axis all the Colours were made to vanifh but the red ; the Light which makes that red being left alone, appeared of the very fame red Colour before its incidence on the third Prifm. And in general we find by other Experiments that when the rays which differ in refrangibility are feparated from one another, and any one fort of them is confidered apart, the Colour of the Light which they compofe cannot be changed by any refraction or reflexion whatever, as it ought to be were Colours nothing elfe than modifications of Light caufed by refractions, and reflexions, and fhadows. This unchangeablenefs of Colour I am now to defcribe in the following Propofition.

PROP. II. THEOR. II.

All homogeneal Light has its proper Colour anfwering to its degree of refrangibility, and that Colour cannot be changed by reflexions and refractions.

In the Experiments of the 4th Propofition of the firft Book, when I had feparated the heterogeneous rays from one another, the Spectrum p t formed by the feparated
rated

rated rays, did in the progrefs from its end p, on which the moft refrangible rays fell, unto its other end t, on which the leaft refrangible rays fell, appear tinged with this Series of Colours, violet, indico, blue, green, yellow, orange, red, together with all their intermediate degrees in a continual fucceffion perpetually varying: So that there appeared as many degrees of Colours, as there were forts of rays differing in refrangibility.

EXPER. V.

Now that thefe Colours could not be changed by refraction, I knew by refracting with a Prifm fometimes one very little part of this Light, fometimes another very little part, as is defcribed in the 12th Experiment of the firft Book. For by this refraction the Colour of the Light was never changed in the leaft. If any part of the red Light was refracted, it remained totally of the fame red Colour as before. No orange, no yellow, no green, or blue, no other new Colour was produced by that refraction. Neither did the Colour any ways change by repeated refractions, but continued always the fame red entirely as at firft. The like conftancy and immutability I found alfo in the blue, green, and other Colours. So alfo if I looked through a Prifm upon any body illuminated with any part of this homogeneal Light, as in the 14th Experiment of the firft Book is defcribed; I could not perceive any new Colour generated this way. All Bodies illuminated with compound Light appear through Prifms confufed (as was faid above) and tinged with various new Colours, but thofe illuminated with homogeneal Light appeared
through

through Prifms neither lefs diftinct, nor otherwife coloured, than when viewed with the naked Eyes. Their
Colours were not in the leaft changed by the refraction
of the interpofed Prifm. I fpeak here of a fenfible
change of Colour : For the Light which I here call homogeneal, being not abfolutely homogeneal, there ought
to arife fome little change of Colour from its heterogeneity. But if that heterogeneity was fo little as it might
be made, by the faid Experiments of the fourth Propofition, that change was not fenfible, and therefore, in
Experiments where fenfe is judge, ought to be accounted none at all.

EXPER. VI.

And as thefe Colours were not changeable by refractions, fo neither were they by reflexions. For all
white, grey, red, yellow, green, blue, violet Bodies, as
Paper, Afhes, red Lead, Orpiment, Indico, Bife, Gold,
Silver, Copper, Grafs, blue Flowers, Violets, Bubbles
of Water tinged with various Colours, Peacock's Feathers, the tincture of *Lignum Nephriticum*, and fuch
like, in red homogeneal Light appeared totally red, in
blue Light totally blue, in green Light totally green,
and fo of other Colours. In the homogeneal Light of
of any Colour they all appeared totally of that fame
Colour, with this only difference, that fome of them
reflected that Light more ftrongly, others more faintly.
I never yet found any Body which by reflecting homogeneal Light could fenfibly change its Colour.

M From

From all which it is manifeft, that if the Sun's Light confifted of but one fort of rays, there would be but one Colour in the whole World, nor would it be poffible to produce any new Colour by reflexions and refractions, and by confequence that the variety of Colours depends upon the compofition of Light.

DEFINITION.

The homogeneal light and rays which appear red, or rather make Objects appear fo, I call rubrific or red-makng ; thofe which make Objects appear yellow, green, blue and violet, I call yellow-making, green-making; blue-making, violet-making, and fo of the reft. And if at any time I fpeak of light and rays as coloured or endued with Colours, I would be underftood to fpeak not philofophically and properly, but grofly, and according to fuch conceptions as vulgar People in feeing all thefe Experiments would be apt to frame. For the rays to fpeak properly are not coloured. In them there is nothing elfe than a certain power and difpofition to ftir up a fenfation of this or that Colour. For as found in a Bell or mufical String, or other founding Body, is nothing but a trembling Motion, and in the Air nothing but that Motion propagated from the Object, and in the Senforium 'tis a fenfe of that Motion under the form of found ; fo Colours in the Object are nothing but a difpofition to reflect this or that fort of rays more copioufly than the reft ; in the rays they are nothing but their difpofitions to propagate
gate

gate this or that Motion into the Senforium, and in the Senforium they are senfations of thofe Motions under the forms of Colours.

PROP. III. PROB. I.

To define the refrangibility of the feveral forts of homogeneal Light anfwering to the feveral Colours.

For determining this Problem I made the following Experiment.

EXPER. VII.

When I had caufed the rectilinear line fides A F, G M, *Fig.* 4. of the Spectrum of Colours made by the Prifm to be diftinctly defined, as in the fifth Experiment of the firft Book is defcribed, there were found in it all the homogeneal Colours in the fame order and fituation one among another as in the Spectrum of fimple Light, defcribed in the fourth Experiment of that Book. For the Circles of which the Spectrum of compound Light PT is compofed, and which in the middle parts of the Spectrum interfere and are intermixt with one another, are not intermixt in their outmoft parts where they touch thofe rectilinear fides A F and G M. And therefore in thofe rectilinear fides when diftinctly defined, there is no new Colour generated by refraction. I obferved alfo, that if any where between the two outmoft Circles T M F and P G A a right line, as γδ, was crofs to the Spectrum, fo as at both ends to fall perpendicularly upon its rectilinear fides, there appeared

one

one and the fame Colour and degree of Colour from one
end of this line to the other. I delineated therefore in
a Paper the perimeter of the Spectrum FAPGMT,
and in trying the third Experiment of the first Book, I
held the Paper fo that the Spectrum might fall upon
this delineated Figure, and agree with it exactly, whilst
an Affiftant whofe Eyes for diftinguifhing Colours were
more critical than mine, did by right lines αβ, γδ, εζ,&c.
drawn crofs the Spectrum, note the confines of the Colours that is of the red MαβF of the orange αγδβ, of
the yellow γεζδ, of the green εηθζ, of the blue ηιϰθ,
of the indico ιλμϰ, and of the violet λGAμ. And
this operation being divers times repeated both in the
fame and in feveral Papers, I found that the Obfervations agreed well enough with one another, and
that the rectilinear fides MG and FA were by the faid
crofs lines divided after the manner of a mufical Chord.
Let GM be produced to X, that MX may be equal
to GM, and conceive GX, λX, ιX, ηX, εX, γX, αX,
MX, to be in proportion to one another, as the num-
bers $1, \frac{8}{9}, \frac{5}{6}, \frac{3}{4}, \frac{2}{3}, \frac{3}{5}, \frac{9}{16}, \frac{1}{2}$, and fo to reprefent the
Chords of the Key, and of a Tone, a third Minor, a
fourth, a fifth, a fixth Major, a feventh, and an eighth
above that Key: And the intervals Mα, αγ, γε, εη, ηι,
ιλ, and λG, will be the fpaces which the feveral Colours (red, orange, yellow, green, blue, indico, violet)
take up.

Now thefe intervals or fpaces fubtending the differences of the refractions of the rays going to the limits
of thofe Colours, that is, to the points M, α, γ, ε, η, ι, λ, G,
may without any fenfible Error be accounted proportional to the differences of the fines of refraction of thofe

rays

rays having one common fine of incidence, and therefore fince the common fine of incidence of the moft and
leaft refrangible rays out of Glafs into Air was, (by a
method defcribed above) found in proportion to their
fines of refraction, as 50 to 77 and 78, divide the difference between the fines of refraction 77 and 78, as the
line G M is divided by thofe intervals, you will have
77, $77\frac{1}{8}$, $77\frac{1}{5}$, $77\frac{1}{3}$, $77\frac{1}{2}$, $77\frac{2}{3}$, $77\frac{7}{9}$, 78, the fines of
refraction of thofe rays out of Glafs into Air, their
common fine of incidence being 50. So then the fines
of the incidences of all the red-making rays out of
Glafs into Air, were to the fines of their refractions,
not greater than 50 to 77, nor lefs than 50 to $77\frac{1}{8}$, but
varied from one another according to all interme-
diate Proportions. And the fines of the incidences
of the green-making rays were to the fines of
their refractions in all proportions from that of 50
to $77\frac{1}{3}$, unto that of 50 to $77\frac{1}{2}$. And by the like limits
above-mentioned were the refractions of the rays be-
longing to the reft of the Colours defined, the fines of
the red-making rays extending from 77 to $77\frac{1}{8}$, thofe
of the orange-making from $77\frac{1}{8}$ to $77\frac{1}{5}$, thofe of the yellow-making from $77\frac{1}{5}$ to $77\frac{1}{3}$, thofe of the green-making
from $77\frac{1}{3}$ to $77\frac{1}{2}$, thofe of the blue-making from $77\frac{1}{2}$ to
$77\frac{2}{3}$, thofe of the indico-making from $77\frac{2}{3}$ to $77\frac{7}{9}$, and
thofe of the violet from $77\frac{7}{9}$ to 78.

Thefe are the Laws of the refractions made out of
Glafs into Air, and thence by the three Axioms of the
firft Book the Laws of the refractions made out of Air
into Glafs are eafily derived.

EXPER.

EXPER. VIII.

I found moreover that when Light goes out of Air through feveral contiguous refracting Mediums as through Water and Glafs, and thence goes out again into Air, whether the refracting fuperficies be parallel or inclined to one another, that Light as often as by contrary refractions 'tis fo corrected, that it emergeth in lines parallel to thofe in which it was incident, continues ever after to be white. But if the emergent rays be inclined to the incident, the whitenefs of the emerging Light will by degrees in paffing on from the place of emergence, become tinged in its edges with Colours. This I tryed by refracting Light with Prifms of Glafs within a prifmatick Veffel of Water. Now thofe Colours argue a diverging and feparation of the heterogeneous rays from one another by means of their unequal refractions, as in what follows will more fully appear. And, on the contrary, the permanent whitenefs argues, that in like incidences of the rays there is no fuch feparation of the emerging rays, and by confequence no inequality of their whole refractions. Whence I feem to gether the two following Theorems.

1. The Exceffes of the fines of refraction of feveral forts of rays above their common fine of incidence when the refractions are made out of divers denfer mediums immediately into one and the fame rarer medium, are to one another in a given Proportion.

2. The

2. The Proportion of the fine of incidence to the fine
of refraction of one and the fame fort of rays out of one
medium into another, is compofed of the Proportion of
the fine of incidence to the fine of refraction out of the
firft medium into any third medium, and of the Proportion of the fine of incidence to the fine of refraction
out of that third medium into the fecond medium.

By the firft Theorem the refractions of the rays of
every fort made out of any medium into Air are known
by having the refraction of the rays of any one fort. As
for inftance, if the refractions of the rays of every fort
out of Rain-water into Air be defired, let the common
fine of incidence out of Glafs into Air be fubducted
from the fines of refraction, and the Exceffes will be
27, $27\frac{1}{8}$, $27\frac{1}{5}$, $27\frac{1}{3}$, $27\frac{1}{2}$, $27\frac{5}{3}$, $27\frac{7}{9}$, 28. Suppofe now
that the fine of incidence of the leaft refrangible rays be
to their fine of refraction out of Rain-water into Air as
three to four, and fay as 1 the difference of thofe fines
is to 3 the fine of incidence, fo is 27 the leaft of the
Exceffes above-mentioned to a fourth number 81 ; and
81 will be the common fign of incidence out of Rain-
water into Air, to which fine if you add all the above-
mentioned Exceffes you will have the defired fines of
the refractions 108, $108\frac{1}{8}$, $108\frac{1}{5}$, $108\frac{1}{3}$, $108\frac{1}{2}$, $108\frac{2}{3}$,
$108\frac{7}{9}$, 109.

By the latter Theorem the refraction out of one medium into another is gathered as often as you have
the refractions out of them both into any third medium.
As if the fine of incidence of any ray out of Glafs into
Air be to its fine of refraction as 20 to 31, and the fine
of incidence of the fame ray out of Air into Water, be

to

to its fine of refraction as four to three; the fine of incidence of that ray out of Glafs into Water will be to its fine of refraction as 20 to 31 and 4 to 3 joyntly, that is, as the Factum of 20 and 4 to the Factum of 31 and 3, or as 80 to 93.

And thefe Theorems being admitted into Opticks, there would be fcope enough of handling that Science voluminoufly after a new manner; not only by teaching thofe things which tend to the perfection of vifion, but alfo by determining mathematically all kinds of Phæno-mena of Colours which could be produced by refra-ctions. For to do this, there is nothing elfe requifite than to find out the feparations of heterogeneous rays, and their various mixtures and proportions in every mixture. By this way of arguing I invented almoft all the Phænomena defcribed in thefe Books, befide fome others lefs neceffary to the Argument; and by the fucceffes I met with in the tryals, I dare promife, that to him who fhall argue truly, and then try all things with good Glaffes and fufficient circumfpection, the expected event will not be wanting. But he is firft to know what Colours will arife from any others mixt in any affigned Proportion.

PROP. IV. THEOR. III.

Colours may be produced by compofition which fhall be like to the Colours of homogeneal Light as to the appearance of Colour, but not as to the immutability of Colour and conftitution of Light. And thofe Colours by how much they are more compounded by fo much are they lefs full and inteufe, and by too much compofition they may be

diluted

diluted aud weakened till they ceafe. There may be alfo Colours produced by compofition, which are not fully like any of the Colours of homogeneal Light.

For a mixture of homogeneal red and yellow compounds an orange, like in appearance of Colour to that orange which in the feries of unmixed prifmatick Colours lies between them; but the Light of one orange is homogeneal as to refrangibility, that of the other is heterogeneal, and the Colour of the one, if viewed through a Prifm, remains unchanged, that of the other is changed and refolved into its component Colours red and yellow. And after the fame manner other neighbouring homogeneal Colours may compound new Colours, like the intermediate homogeneal ones, as yellow and green, the Colour between them both, and afterwards, if blue be added, there will be made a green the middle Colour of the three which enter the compofition. For the yellow and blue on either hand, if they are equal in quantity they draw the intermediate green equally towards themfelves in compofition, and fo keep it as it were in equilibrio, that it verge not more to the yellow on the one hand, than to the blue on the other, but by their mixt actions remain ftill a middle Colour. To this mixed green there may be further added fome red and violet, and yet the green will not prefently ceafe but only grow lefs full and vivid, and by increafing the red and violet it will grow more and more dilute, until by the prevalence of the added Colours it be overcome and turned into whitenefs, or fome other Colour. So if to the Colour of any homogeneal Light, the Sun's white Light compofed of all forts of rays be

N   added,

added, that Colour will not vanish or change its species but be diluted, and by adding more and more white it will be diluted more and more perpetually. Lastly, if red and violet be mingled, there will be generated according to their various Proportions various Purples, such as are not like in appearance to the Colour of any homogeneal Light, and of these Purples mixt with yellow and blue may be made other new Colours.

PROP. V. THEOR. IV.

Whiteness and all grey Colours between white and black, may be compounded of Colours, and the whiteness of the Sun's Light is compounded of all the primary Colours mixt in a due proportion.

The Proof by Experiments.

EXPER. IX.

Fig. 5. The Sun shining into a dark Chamber through a little round Hole in the Window shut, and his Light being there refracted by a Prism to cast his coloured Image PT upon the opposite Wall : I held a white Paper V to that Image in such manner that it might be illuminated by the coloured Light reflected from thence, and yet not intercept any part of that Light in its passage from the Prism to the Spectrum. And I found that when the Paper was held nearer to any Colour than to the rest, it appeared of that Colour to which it approached nearest ; but when it was equally or almost

equally

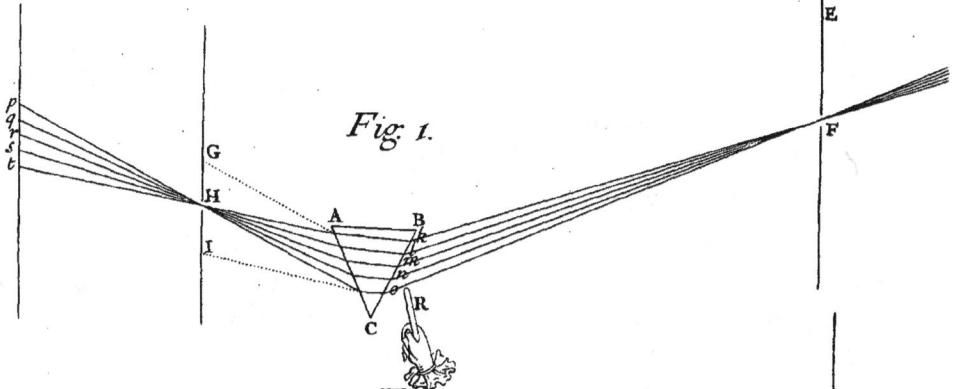

Fig. 1.

Fig. 2.

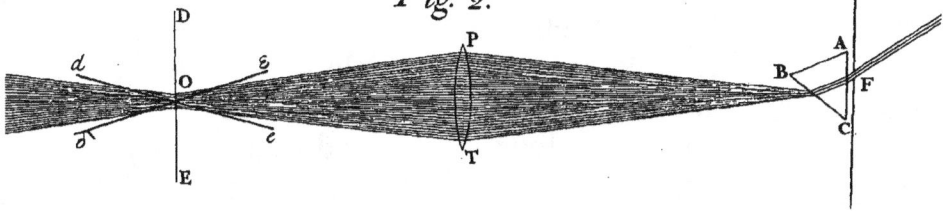

Fig. 3.

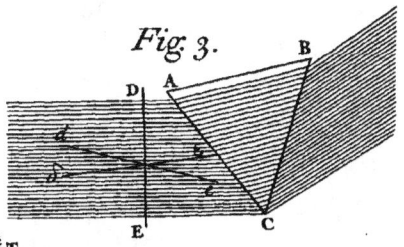

Fig. 4.

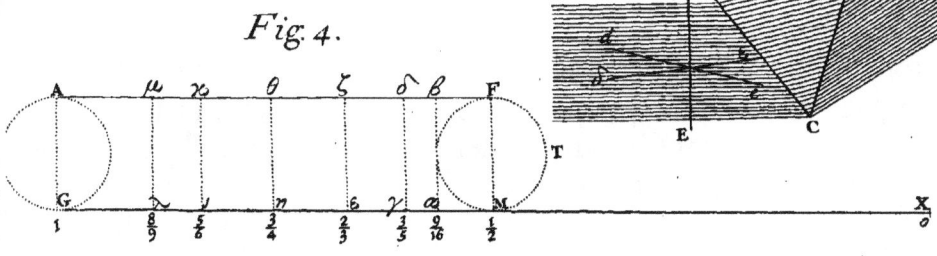

Fig. 5.

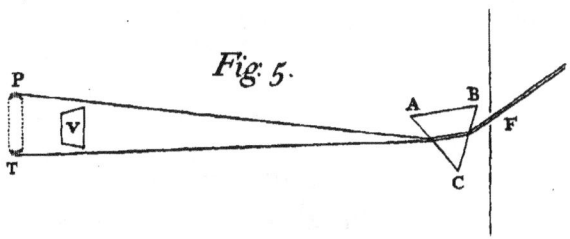

equally diftant from all the Colours, fo that it might be equally illuminated by them all it appeared white. And in this laft fituation of the Paper, if fome Colours were intercepted, the Paper loft its white Colour, and appeared of the Colour of the reft of the Light which was not intercepted. So then the Paper was illuminated with Lights of various Colours, namely, red, yellow, green, blue and violet, and every part of the Light retained its proper Colour, until it was incident on the Paper, and became reflected thence to the Eye ; fo that if it had been either alone (the reft of the Light being intercepted) or if it had abounded moft and been predominant in the Light reflected from the Paper, it would have tinged the Paper with its own Colour; and yet being mixed with the reft of the Colours in a due proportion, it made the Paper look white, and therefore by a compofition with the reft produced that Colour. The feveral parts of the coloured Light reflected from the Spectrum, whilft they are propagated from thence thro' the Air, do perpetually retain their proper Colours, becaufe wherever they fall upon the Eyes of any Spectator, they make the feveral parts of the Spectrum to appear under their proper Colours. They retain therefore their proper Colours when they fall upon the Paper V, and fo by the confufion and perfect mixture of thofe Colours compound the whitenefs of the Light reflected from thence.

EXPER. X.

Let that Spectrum or folar Image P T fall now upon *Fig.* 6. the Lens M N above four Inches broad, and about fix

Feet

Feet diſtant from the Priſm A B C, and ſo figured that
it may cauſe the coloured Light which divergeth from
the Priſm to converge and meet again at its Focus G,
about ſix or eight Feet diſtant from the Lens, and
there to fall perpendicularly upon a white Paper D E.
And if you move this Paper to and fro, you will perceive that near the Lens, as at *de*, the whole ſolar Image
(ſuppoſe at p t) will appear upon it intenſly coloured
after the manner above-explained, and that by receding
from the Lens thoſe Colours will perpetually come towards one another, and by mixing more and more dilute one another continually, until at length the Paper
come to the Focus G, where by a perfect mixture they
will wholly vaniſh and be converted into whiteneſs, the
whole Light appearing now upon the Paper like a little
white Circle. And afterwards by receding further from
the Lens, the rays which before converged will now
croſs one another in the Focus G, and diverge from
thence, and thereby make the Colours to appear again,
but yet in a contrary order ; ſuppoſe at *de*, where the
red t is now above which before was below, and the
violet p is below which before was above.

Let us now ſtop the Paper at the Focus G where
the Light appears totally white and circular, and let us
conſider its whiteneſs. I ſay, that this is compoſed of
the converging Colours. For if any of thoſe Colours
be intercepted at the Lens, the whiteneſs will ceaſe and
degenerate into that Colour which ariſeth from the
compoſition of the other Colours which are not intercepted. And then if the intercepted Colours be let
paſs and fall upon that compound Colour, they mix
with it, and by their mixture reſtore the whiteneſs.

So

So if the violet, blue and green be intercepted, the remaining yellow, orange and red will compound upon the Paper an orange, and then if the intercepted Colours be let pass they will fall upon this compounded orange, and together with it decompound a white. So also if the red and violet be intercepted, the remaining yellow, green and blue, will compound a green upon the Paper, and then the red and violet being let pass will fall upon this green, and together with it decompound a white. And that in this composition of white the several rays do not suffer any change in their colorific qualities by acting upon one another, but are only mixed, and by a mixture of their Colours produce white, may further appear by these Arguments.

If the Paper be placed beyond the Focus G, suppose at *δε*, and then the red Colour at the Lens be alternately intercepted, and let pass again, the violet Colour on the Paper will not suffer any change thereby, as it ought to do if the several sorts of rays acted upon one another in the Focus G, where they cross. Neither will the red upon the Paper be changed by any alternate stopping, and letting pass the violet which crosseth it.

And if the Paper be placed at the Focus G, and the white round Image at G be viewed through the Prism H1K, and by the refraction of that Prism be translated to the place r v, and there appear tinged with various Colours, namely, the violet at v and red at r, and others between, and then the red Colour at the Lens be often stopt and let pass by turns, the red at r will accordingly disappear and return as often, but the violet at v will not thereby suffer any change. And so by stopping and letting pass alternately the blue at the
Lens,

Lens, the blue at r will accordingly difappear and return, without any change made in the red at r. The
red therefore depends on one fort of rays, and the blue
on another fort, which in the Focus G where they are
commixt do not act on one another. And there is the
fame reafon of the other Colours.

I confidered further, that when the moft refrangible
rays P p, and the leaft refrangible ones T t, are by converging inclined to one another, the Paper, if held very
oblique to thofe rays in the Focus G, might reflect one
fort of them more copioufly than the other fort, and by
that means the reflected Light would be tinged in that
Focus with the Colour of the predominant rays, provided thofe rays feverally retained their Colours or colorific qualities in the compofition of white made by
them in that Focus. But if they did not retain them
in that white, but became all of them feverally endued
there with a difpofition to ftrike the fenfe with the per-
ception of white, then they could never lofe their whitenefs by fuch reflexions. I inclined therefore the Paper
to the rays very obliquely, as in the fecond Experiment
of this Book, that the moft refrangible rays might be
more copioufly reflected than the reft, and the whitenefs at length changed fucceffively into blue, indico
and violet. Then I inclined it the contrary way, that
the moft refrangible rays might be more copious in the
reflected Light than the reft, and the whitenefs turned
fucceffively to yellow, orange and red.

Laftly, I made an Inftrument X Y in fafhion of a
Comb, whofe Teeth being in number fixteen were
about an Inch and an half broad, and the intervals of the
Teeth about two Inches wide. Then by interpofing
fuc-

fucceffively the Teeth of this Inftrument near the Lens,
I intercepted part of the Colours by the interpofed
Tooth, whilft the reft of them went on through the interval of the Teeth to the Paper D E, and there painted a round folar Image. But the Paper I had firft placed fo, that the Image might appear white as often
as the Comb was taken away; and then the Comb being as was faid interpofed, that whitenefs by reafon of
the intercepted part of the Colours at the Lens did always change into the Colour compounded of thofe
Colours which were not intercepted, and that Colour
was by the motion of the Comb perpetually varied fo,
that in the paffing of every Tooth over the Lens all
thefe Colours red, yellow, green, blue and purple, did
always fucceed one another. I caufed therefore all the
Teeth to pafs fucceffively over the Lens, and when the
motion was flow, there appeared a perpetual fucceffion
of the Colours upon the Paper: But if I fo much accele-
lerated the motion, that the Colours by reafon of their
quick fucceffion could not be diftinguifhed from one
another, the appearance of the fingle Colours ceafed.
There was no red, no yellow, no green, no blue, nor
purple to be feen any longer, but from a confufion of
them all there arofe one uniform white Colour. Of the
Light which now by the mixture of all the Colours appeared white, there was no part really white. One
part was red, another yellow, a third green, a fourth
blue, a fifth purple, and every part retains its proper
Colour till it ftrike the Senforium. If the impreffions
follow one another flowly, fo that they may be feverally perceived, there is made a diftinct fenfation of all
the Colours one after another in a continual fucceffion.

But

But if the impreſſions follow one another ſo quickly that they cannot be ſeverally perceived, there ariſeth out of them all one common ſenſation, which is neither of this Colour alone nor of that alone, but hath it ſelf indifferently to 'em all, and this is a ſenſation of whiteneſs. By the quickneſs of the ſucceſſions the impreſſions of the ſeveral Colours are confounded in the Senſorium, and out of that confuſion ariſeth a mixt ſenſation. If a burning Coal be nimbly moved round in a Circle with Gyrations continually repeated, the whole Circle will appear like fire ; the reaſon of which is, that the ſenſation of the Coal in the ſeveral places of that Circle remains impreſt on the Senſorium, until the Coal return again to the ſame place. And ſo in a quick confecution of the Colours the impreſſion of every Colour remains in the Senſorium, until a revolution of all the Colours be compleated, and that firſt Colour return again. The impreſſions therefore of all the ſucceſſive Colours are at once in the Senſorium, and joyntly ſtir up a ſenſation of them all ; and ſo it is manifeſt by this Experiment, that the commixt impreſſions of all the Colours do ſtir up and beget a ſenſation of white, that is, that whiteneſs is compounded of all the Colours.

And if the Comb be now taken away, that all the Colours may at once paſs from the Lens to the Paper, and be there intermixed, and together reflected thence to the Spectators Eyes ; their impreſſions on the Senſorium being now more ſubtily and perfectly commixed there, ought much more to ſtir up a ſenſation of whiteneſs.

You

You may inftead of the Lens ufe two Prifms H I K and L M N, which by refracting the coloured Light the contrary way to that of the firft refraction, may make the diverging rays converge and meet again in G, as you fee it reprefented in the feventh Figure. For *Fig.* 7. where they meet and mix they will compofe a white Light as when a Lens is ufed.

E X P E R. XI.

Let the Sun's coloured Image P T fall upon the Wall *Fig.* 8. of a dark Chamber, as in the third Experiment of the firft Book, and let the fame be viewed through a Prifin a b c, held parallel to the Prifm A B C, by whofe refraction that Image was made, and let it now appear lower than before, fuppofe in the place S over againft the red colour T. And if you go near to the Image P T, the Spectrum S will appear oblong and coloured like the Image P T ; but if you recede from it, the Colours of the Spectrum S will be contracted more and more, and at length vanifh, that Spectrum S becoming perfectly round and white ; and if you recede yet further, the Colours will emerge again, but in a contrary order. Now that Spectrum S appears white in that cafe when the rays of feveral forts which converge from the feveral parts of the Image P T, to the Prifin a b c, are fo refracted unequally by it, that in their paffage from the Prifin to the Eye they may diverge from one and the fame point of the Spectrum S, and fo fall afterwards upon one and the fame point in the bottom of the Eye, and there be mingled.

And

And further, if the Comb be here made ufe of, by whofe Teeth the Colours at the Image PT may be fucceffively intercepted ; the Spectrum S when the Comb is moved flowly will be perpetually tinged with fucceffive Colours: But when by accelerating the motion of the Comb, the fucceffion of the Colours is fo quick that they cannot be feverally feen, that Spectrum S, by a confufed and mixt fenfation of them all, will appear white.

EXPER. XII.

Fig. 9.

The Sun fhining through a large Prifm A BC upon a Comb X Y, placed immediately behind the Prifm, his Light which paffed through the interftices of the Teeth fell upon a white Paper DE. The breadths of the Teeth were equal to their interftices, and feven Teeth together with their interftices took up an Inch in breadth. Now when the Paper was about two or three Inches diftant from the Comb, the Light which paffed through its feveral interftices painted fo many ranges of Colours kl, mn, op, qr, &c. which were parallel to one another and contiguous, and without any mixture of white. And thefe ranges of Colours, if the Comb was moved continually up and down with a reciprocal motion, afcended and defcended in the Paper, and when the motion of the Comb was fo quick, that the Colours could not be diftinguifhed from one another, the whole Paper by their confufion and mixture in the Senforium appeared white.

Let

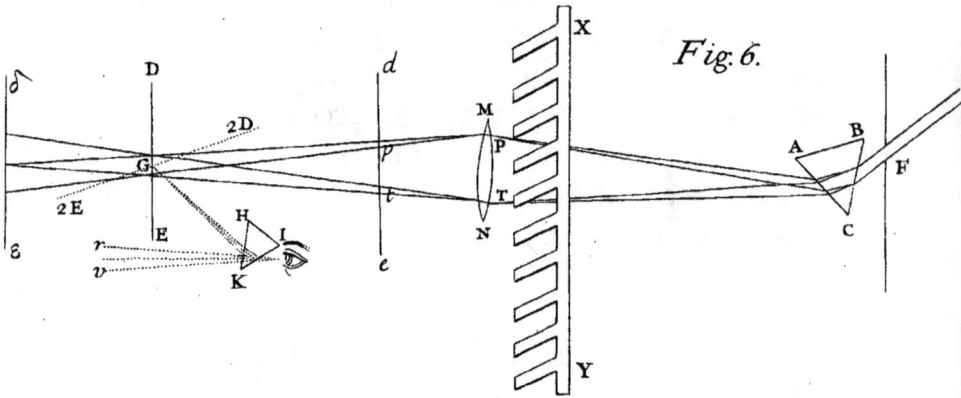

Fig. 6.

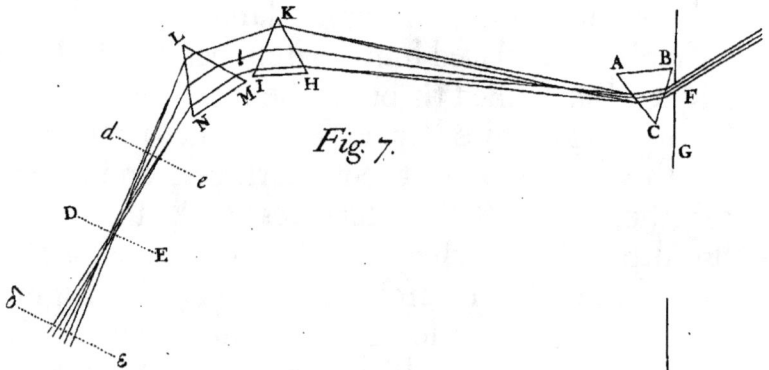

Fig. 7.

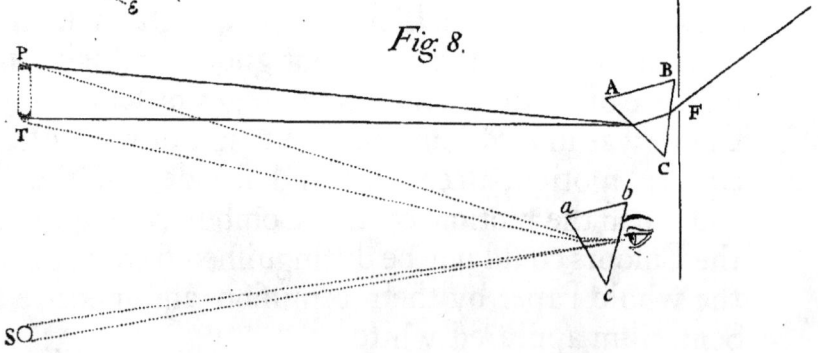

Fig. 8.

Let the Comb now reft, and let the Paper be removed further from the Prifm, and the feveral ranges of Colours will be dilated and expanded into one another more and more, and by mixing their Colours will dilute one another, and at length, when the diftance of the Paper from the Comb is about a Foot, or a little more (fuppofe in the place 2 D 2 E) they will fo far dilute one another as to become white.

With any Obftacle let all the Light be now ftopt which paffes through any one interval of the Teeth, fo that the range of Colours which comes from thence may be taken away, and you will fee the Light of the reft of the ranges to be expanded into the place of the range taken away, and there to be coloured. Let the intercepted range pafs on as before, and its Colours falling upon the Colours of the other ranges, and mixing with them, will reftore the whitenefs.

Let the Paper 2 D 2 E be now very much inclined to the rays, fo that the moft refrangible rays may be more copioufly reflected than the reft, and the white Colour of the Paper through the excefs of thofe rays will be changed into blue and violet. Let the Paper be as much inclined the contrary way, that the leaft refrangible rays may be now more copioufly reflected than the reft, and by their excefs the whitenefs will be changed into yellow and red. The feveral rays therefore in that white Light do retain their colorific qualities, by which thofe of any fort, when-ever they become more copious than the reft, do by their excefs and predominance caufe their proper Colour to appear.

And

And by the fame way of arguing, applied to the third
Experiment of this Book, it may be concluded, that
the white Colour of all refracted Light at its very firft
emergence, where it appears as white as before its incidence, is compounded of various Colours.

EXPER. XIII.

In the foregoing Experiment the feveral intervals of
the Teeth of the Comb do the office of fo many Prifms,
every interval producing the Phænomenon of one Prifm.
Whence inftead of thofe intervals ufing feveral Prifms, I
try'd to compound whitenefs by mixing their Colours, and
did it by ufing only three Prifms, as alfo by ufing only
Fig. 10. two as follows. Let two Prifms A B C and a b c, whofe
refracting Angles B and b are equal, be fo placed parallel
to one another, that the refracting Angle B of the one
may touch the Angle c at the bafe of the other, and
their planes C B and c b, at which the rays emerge, may
lye in directum. Then let the Light trajected through
them fall upon the Paper M N, diftant about 8 or 12
Inches from the Prifms. And the Colours generated
by the interior limits B and c of the two Prifms, will
be mingled at P T, and there compound white. For if
either Prifm be taken away, the Colours made by the
other will appear in that place P T, and when the Prifm
is reftored to its place again, fo that its Colours may
there fall upon the Colours of the other, the mixture
of them both will reftore the whitenefs.

This

This Experiment fucceeds alfo, as I have tryed, when the Angle b of the lower Prifm, is a little greater than the Angle B of the upper, and between the interior Angles B and c, there intercedes fome fpace B e, as is reprefented in the Figure, and the refracting planes BC and b c, are neither in directum, nor parallel to one another. For there is nothing more requifite to the fuccefs of this Experiment, than that the rays of all forts may be uniformly mixed upon the Paper in the place P T. If the moft refrangible rays coming from the fuperior Prifm take up all the fpace from M to P, the rays of the fame fort which come from the inferior Prifm ought to begin at P, and take up all the reft of the fpace from thence towards N. If the leaft refrangible rays coming from the fuperior Prifm take up the fpace M T, the rays of the fame kind which come from the other Prifm ought to begin at T, and take up the remaining fpace T N. If one fort of the rays which have intermediate degrees of refrangibility, and come from the fuperior Prifm be extended through the fpace M Q, and another fort of thofe rays through the fpace M R, and a third fort of them through the fpace M S, the fame forts of rays coming from the lower Prifm, ought to illuminate the remaining fpaces Q N, R N, S N refpectively. And the fame is to be underftood of all the other forts of rays. For thus the rays of every fort will be fcattered uniformly and evenly through the whole fpace M N, and fo being every where mixt in the fame proportion, they muft every where produce the fame Colour. And therefore fince by this mixture they produce white in the exterior fpaces M P and T N, they muft alfo produce white in the interior fpace P T. This

is

is the reaſon of the compoſition by which whiteneſs was produced in this Experiment, and by what other way ſoever I made the like compoſition the reſult was whiteneſs.

Laſtly, If with the Teeth of a Comb of a due ſize, the coloured Lights of the two Priſms which fall upon the ſpace PT be alternately intercepted, that ſpace PT, when the motion of the Comb is ſlow, will always appear coloured, but by accelerating the motion of the Comb ſo much, that the ſucceſſive Colours cannot be diſtinguiſhed from one another, it will appear white.

EXPER. XIV.

Hitherto I have produced whiteneſs by mixing the Colours of Priſms. If now the Colours of natural Bodies are to be mingled, let Water a little thickned with Soap be agitated to raiſe a froth, and after that froth has ſtood a little, there will appear to one that ſhall view it intently various Colours every where in the ſurfaces of the ſeveral Bubbles; but to one that ſhall go ſo far off that he cannot diſtinguiſh the Colours from one another, the whole froth will grow white with a perfect whiteneſs.

EXPER. XV.

Laſtly, in attempting to compound a white by mixing the coloured Powders which Painters uſe, I conſidered that all coloured Powders do ſuppreſs and ſtop in them a very conſiderable part of the Light by which

they

they are illuminated. For they become coloured by reflecting the Light of their own Colours more copiously, and that of all other Colours more sparingly, and yet they do not reflect the Light of their own Colours so copiously as white Bodies do. If red Lead, for instance, and a white Paper, be placed in the red Light of the coloured Spectrum made in a dark Chamber by the refraction of a Prism, as is described in the third Eperiment of the first Book ; the Paper will appear more lucid than the red Lead, and therefore reflects the red-making rays more copiously than red Lead doth. And if they be held in the Light of any other Colour, the Light reflected by the Paper will exceed the Light reflected by the red Lead in a much greater proportion. And the like happens in Powders of other Colours. And therefore by mixing such Powders we are not to expect a strong and full white, such as is that of Paper, but some dusky obscure one, such as might arise from a mixture of light and darkness, or from white and black, that is, a grey, or dun, or russet brown, such as are the Colours of a Man's Nail, of a Mouse, of Ashes, of ordinary Stones, of Mortar, of Dust and Dirt in Highways, and the like. And such a dark white I have often produced by mixing coloured Powders. For thus one part of red Lead, and five parts of *Viride Æris*, composed a dun Colour like that of a Mouse. For these two Colours were severally so compounded of others, that in both together were a mixture of all Colours ; and there was less red Lead used than *Viride Æris*, because of the fulness of its Colour. Again, one part of red Lead, and four parts of blue Bise, composed a dun Colour verging a little to purple, and by adding to this a

certain

certain mixture of Orpiment and *Viridi Æris* in a due proportion, the mixture loft its purple tincture, and became perfectly dun. But the Experiment fucceeded beft without Minium thus. To Orpiment I added by little and little a certain full bright purple, which Painters ufe until the Orpiment ceafed to be yellow, and became of a pale red. Then I diluted that red by adding a little *Viride Æris*, and a little more blue Bife than *Viridi Æris*, until it became of fuch a grey or pale white, as verged to no one of the Colours more than to another. For thus it became of a Colour equal in whitenefs to that of Afhes or of Wood newly cut, or of a Man's Skin. The Orpiment reflected more Light than did any other of the Powders, and therefore conduced more to the whitenefs of the compounded Colour than they. To affign the proportions accurately may be difficult, by reafon of the different goodnefs of Powders of the fame kind. Accordingly as the Colour of any Powder is more or lefs full and luminous, it ought to be ufed in a lefs or greater proportion.

Now confidering that thefe grey and dun Colours may be alfo produced by mixing whites and blacks, and by confequence differ from perfect whites not in Species of Colours but only in degree of luminoufnefs, it is manifeft that there is nothing more requifite to make them perfectly white than to increafe their Light fufficiently ; and, on the contrary, if by increafing their Light they can be brought to perfect whitenefs, it will thence alfo follow, that they are of the fame Species of Colour with the beft whites, and differ from them only in the quantity of Light. And this I tryed as follows. I took the third of the above-mentioned grey mixtures

(that

(that which was compounded of Orpiment, Purple, Bife and *Viride Æris)* and rubbed it thickly upon the floor of my Chamber, where the Sun fhone upon it through the opened Cafement ; and by it, in the fha-dow, I laid a piece of white Paper of the fame bignefs. Then going from them to the diftance of 12 or 18 Feet, fo that I could not difcern the unevennefs of the furface of the Powder, nor the little fhadows let fall from the gritty particles thereof ; the Powder appeared intenfly white, fo as to tranfcend even the Paper it felf in white-nefs, efpecially if the Paper were a little fhaded from the Light of the Clouds, and then the Paper compared with the Powder appeared of fuch a grey Colour as the Powder had done before. But by laying the Paper where the Sun fhines through the Glafs of the Window, or by fhutting the Window that the Sun might fhine through the Glafs upon the Powder, and by fuch other fit means of increafing or decreafing the Lights where-with the Powder and Paper were illuminated , the Light wherewith the Powder is illuminated may be made ftronger in fuch a due proportion than the Light wherewith the Paper is illuminated, that they fhall both appear exactly alike in whitenefs. For when I was trying this, a Friend coming to vifit me, I ftopt him at the door, and before I told him what the Colours were, or what I was doing ; I askt him, Which of the two whites were the beft, and wherein they differed ? And after he had at that diftance viewed them well, he anfwered, That they were both good whites, and that he could not fay which was beft, nor wherein their Co-lours differed. Now if you confider, that this white of the Powder in the Sun-fhine was compounded of the

P   Colours

Colours which the component Powders (Orpiment, Purple, Bife, and *Viride Æris*) have in the fame Sun-fhine, you muft acknowledge by this Experiment, as well as by the former, that perfect whitenefs may be compounded of Colours.

From what has been faid it is alfo evident, that the whitenefs of the Sun's Light is compounded of all the Colours wherewith the feveral forts of rays whereof that Light confifts, when by their feveral refrangibili-ties they are feparated from one another, do tinge Paper or any other white Body whereon they fall.   For thofe Colours by Prop. 2. are unchangeable, and whenever all thofe rays with thofe their Colours are mixt again, they reproduce the fame white Light as before.

PROP. VI. PROB. II.

In a mixture of primary Colours, the quantity and quality of each being given, to know the Colour of the compound.

Fig. 11.   With the Center O and Radius O D defcribe a Circle A D F, and diftinguifh its circumference into feven parts D E, E F, F G, G A, A B, B C, C D, proportional to the feven mufical Tones or Intervals of the eight Sounds, *Sol, la, fa, fol, la, mi, fa, fol*, contained in an Eight, that is, proportional to the numbers $\frac{1}{9}, \frac{1}{16}, \frac{1}{10}, \frac{1}{9}, \frac{1}{10}, \frac{1}{16}, \frac{1}{9}$. Let the firft part D E reprefent a red Colour, the fecond E F orange, the third F G yellow, the fourth G H green, the fifth A B blue, the fixth B C indico, and the feventh C D violet.   And conceive that thefe are all the Colours of uncompounded Light gradually

passing

paſſing into one another, as they do when made by Priſms; the circumference D E F G A B C D, repreſenting the whole ſeries of Colours from one end of the Sun's coloured Image to the other, ſo that from D to E be all degrees of red, at E the mean Colour between red and orange, from E to F all degrees of orange, at F the mean between orange and yellow, from F to G all degrees of yellow, and ſo on. Let p be the center of gravity of the Arch D E, and q, r, s, t, v, x, the centers of gravity of the Arches E F, F G, G A, A B, B C and C D reſpectively, and about thoſe centers of gravity let Circles proportional to the number of rays of each Colour in the given mixture be deſcribed; that is, the circle p proportional to the number of the red-making rays in the mixture, the Circle q proportional to the number of the orange-making rays in the mixture, and ſo of the reſt. Find the common center of gravity of all thoſe Circles p, q, r, s, t, v, x. Let that center be Z; and from the center of the Circle A D F, through Z to the circumference, drawing the right line O Y, the place of the point Y in the circumference ſhall ſhew the Colour ariſing from the compoſition of all the Colours in the given mixture, and the line O Z ſhall be proportional to the fulneſs or intenſeneſs of the Colour, that is, to its diſtance from whiteneſs. As if Y fall in the middle between F and G, the compounded Colour ſhall be the beſt yellow; if Y verge from the middle towards F or G, the compounded Colour ſhall accordingly be a yellow, verging towards orange or green. If Z fall upon the circumference the Colour ſhall be intenſe and florid in the higheſt degree; if it fall in the mid way between the circumference and center, it ſhall be

P 2

but

but half so intense, that is, it shall be such a Colour as would be made by diluting the intensest yellow with an equal quantity of whiteness; and if it fall upon the center O, the Colour shall have lost all its intenseness, and become a white. But it is to be noted, That if the point Z fall in or near the line O D, the main ingredients being the red and violet, the Colour compounded shall not be any of the prismatic Colours, but a purple, inclining to red or violet, accordingly as the point Z lieth on the side of the line DO towards E or towards C, and in general the compounded violet is more bright and more fiery than the uncompounded. Also if only two of the primary Colours which in the Circle are opposite to one another be mixed in an equal proportion, the point Z shall fall upon the center O, and yet the Colour compounded of those two shall not be perfectly white, but some faint anonymous Colour. For I could never yet by mixing only two primary Colours produce a perfect white. Whether it may be compounded of a mixture of three taken at equal distances in the circumference I do not know, but of four or five I do not much question but it may. But these are curiosities of little or no moment to the understanding the Phænomena of nature. For in all whites produced by nature, there uses to be a mixture of all forts of rays, and by consequence a composition of all Colours.

To give an instance of this Rule; suppose a Colour is compounded of these homogeneal Colours, of violet 1 part, of indico 1 part, of blue 2 parts, of green 3 parts, of yellow 5 parts, of orange 6 parts, and of red 10 parts. Proportional to these parts I describe the Circles x, v, t, s, r, q, p respectively, that is, so that if the Circle x

be

be 1, the Circle v may be 1, the Circle t 2, the Circle
s 3, and the Circles r, q and p, 5, 6 and 10. Then I
find Z the common center of gravity of thefe Circles,
and through Z drawing the line O Y, the point Y falls
upon the circumference between E and F, fome thing
nearer to E than to F, and thence I conclude, that the
Colour compounded of thefe ingredients will be an
orange, verging a little more to red than to yellow.
Alfo I find that O Z is a little lefs than one half of
O Y, and thence I conclude, that this orange hath a
little lefs than half the fulnefs or intenfenefs of an uncompounded orange; that is to fay, that it is fuch an
orange as may be made by mixing an homogeneal orange
with a good white in the proportion of the line O Z to
the line Z Y, this proportion being not of the quantities
of mixed orange and white powders, but of the quantities of the lights reflected from them.

This Rule I conceive accurate enough for practife,
though not mathematically accurate ; and the truth of
it may be fufficiently proved to fenfe, by ftopping any
of the Colours at the Lens in the tenth Experiment of
this Book. For the reft of the Colours which are not
ftopped, but pafs on to the Focus of the Lens, will
there compound either accurately or very nearly fuch
a Colour as by this Rule ought to refult from their
mixture.

PROP.

PROP. VII. THEOR. V.

*All the Colours in the Univerfe which are made by Light,
and depend not on the power of imagination, are
either the Colours of homogeneal Lights, or compounded
of thefe and that either accurately or very nearly, according to the Rule of the foregoing Problem.*

For it has been proved (in Prop.1. *Lib.*2.) that the
changes of Colours made by refractions do not arife
from any new modifications of the rays impreft by thofe
refractions, and by the various terminations of light
and fhadow, as has been the conftant and general opinion of Philofophers. It has alfo been proved that the
feveral Colours of the homogeneal rays do conftantly
anfwer to their degrees of refrangibility, (Prop.1.*Lib.*1.
and Prop.2. *Lib.*2.) and that their degrees of refrangibility cannot be changed by refractions and reflexions,
(Prop.2. *Lib.*1.) and by confequence that thofe their
Colours are likewife immutable. It has alfo been proved directly by refracting and reflecting homogeneal
Lights apart, that their Colours cannot be changed,
(Prop.2. *Lib.*2.) It has been proved alfo, that when
the feveral forts of rays are mixed, and in croffing pafs
through the fame fpace, they do not act on one another
fo as to change each others colorifick qualities, (Exper.
10. *Lib.*2.) but by mixing their actions in the Senforium beget a fenfation differing from what either would
do apart, that is a fenfation of a mean Colour between
their proper Colours ; and particularly when by the
concourfe and mixtures of all forts of rays, a white
Colour

Colour is produced, the white is a mixture of all the Colours which the rays would have apart, (Prop. 5. *Lib.* 2.) The rays in that mixture do not lose or alter their several colorifick qualities, but by all their various kinds of actions mixt in the Senforium, beget a fenfation of a middling Colour between all their Colours which is whitenefs. For whitenefs is a mean between all Colours, having it felf indifferently to them all, fo as with equal facility to be tinged with any of them. A red Powder mixed with a little blue, or a blue with a little red, doth not prefently lofe its Colour, but a white Powder mixed with any Colour is prefently tinged with that Colour, and is equally capable of being tinged with any Colour what-ever. It has been fhewed alfo, that as the Sun's Light is mixed of all forts of rays, fo its whitenefs is a mixture of the Colours of all forts of rays ; thofe rays having from the beginning their feveral colorific qualities as well as their feveral refrangibilities, and retaining them perpetually unchang'd notwithstanding any refractions or reflexions they may at any time fuffer, and that when-ever any fort of the Sun's rays is by any means (as by reflexion in Exper. 9 and 10. *Lib.* 1. or by refraction as happens in all refractions) feparated from the reft, they then manifeft their proper Colours. Thefe things have been proved, and the fum of all this amounts to the Propofition here to be proved. For if the Sun's Light is mixed of feveral forts of rays, each of which have originally their feveral refrangibilities and colorifick qualities, and notwithstanding their refractions and reflections, and their various feparations or mixtures, keep thofe their original properties perpetually the fame without alteration ;

tion ; then all the Colours in the World muſt be ſuch as
conſtantly ought to ariſe from the original colorific qualities of the rays whereof the Lights conſiſt by which
thoſe Colours are ſeen. And therefore if the reaſon of
any Colour what-ever be required, we have nothing elſe
to do then to conſider how the rays in the Sun's Light
have by reflexions or refractions, or other cauſes been parted from one another, or mixed together ; or otherwiſe to
find out what ſorts of rays are in the Light by which
that Colour is made, and in what proportion ; and
then by the laſt Problem to learn the Colour which
ought to ariſe by mixing thoſe rays (or their Colours)
in that proportion. I ſpeak here of Colours ſo far as
they ariſe from Light. For they appear ſometimes by
other cauſes, as when by the power of phantaſy we
ſee Colours in a Dream, or a mad Man ſees things before
him which are not there ; or when we ſee Fire by ſtriking
the Eye, or ſee Colours like the Eye of a Peacock's
Feather, by preſſing our Eyes in either corner whilſt
we look the other way. Where theſe and ſuch like
cauſes interpoſe not, the Colour always anſwers to
the ſort or ſorts of the rays whereof the Light conſiſts,
as I have conſtantly found in what-ever Phænomena of
Colours I have hitherto been able to examin. I ſhall in
the following Propoſitions give inſtances of this in the
Phænomena of chiefeſt note.

PROP.

PROP. VIII. PROB. III.

By the discovered Properties of Light to explain the Colours made by Prisms.

Let A B C represent a Prism refracting the Light of Fig. 12. the Sun, which comes into a dark Chamber through a Hole F φ almost as broad as the Prism, and let M N represent a white Paper on which the refracted Light is cast, and suppose the most refrangible or deepest violet making rays fall upon the space P π, the least refrangible or deepest red-making rays upon the space T γ, the middle sort between the Indico-making and blue-making rays upon the space Q χ, the middle sort of the green-making rays upon the space R ε, the middle sort between the yellow-making and orange-making rays upon the space S σ, and other intermediate sorts upon intermediate spaces. For so the spaces upon which the several sorts adequately fall will by reason of the different refrangibility of those sorts be one lower than another. Now if the Paper MN be so near the Prism that the spaces P T and πγ do not interfere with one another, the distance between them T π will be illuminated by all the sorts of rays in that proportion to one another which they have at their very first coming out of the Prism, and consequently be white. But the spaces PT and πγ on either hand, will not be illuminated by them all, and therefore will appear coloured. And particularly at P, where the outmost violet-making rays fall alone, the Colour must be the deepest violet. At Q where the violet-making and indico-making rays are mixed, it

Q

must

muſt be a violet inclining much to indico. At R where the violet-making, indico-making, blue-making, and one half of the green-making rays are mixed, their Colours muſt (by the conſtruction of the ſecond Problem) compound a middle Colour between indico and blue. At S where all the rays are mixed except the red-making and orange-making, their Colours ought by the ſame Rule to compound a faint blue, verging more to green than indie. And in the progreſs from S to T, this blue will grow more and more faint and dilute, till at T, where all the Colours begin to be mixed, it end in whiteneſs.

So again, on the other ſide of the white at T, where the leaſt refrangible or utmoſt red-making rays are alone the Colour muſt be the deepeſt red. At σ the mixture of red and orange will compound a red inclining to orange. At ε the mixture of red, orange, yellow, and one half of the green muſt compound a middle Colour between orange and yellow. At χ the mixture of all Colours but violet and indico will compound a faint yellow, verging more to green than to orange. And this yellow will grow more faint and dilute continually in its progreſs from χ to π, where by a mixture of all ſorts of rays it will become white.

Theſe Colours ought to appear were the Sun's Light perfectly white: But becauſe it inclines to yellow, the exceſs of the yellow-making rays whereby 'tis tinged with that Colour, being mixed with the faint blue between S and T, will draw it to a faint green. And ſo the Colours in order from P to T ought to be violet, indico, blue, very faint green, white, faint yellow, orange, red. Thus it is by the computation : And they that pleaſe to

view

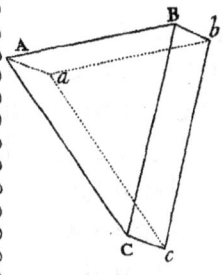

Fig. 9.

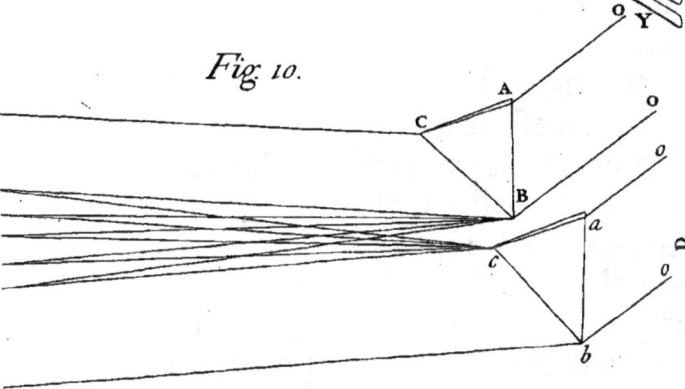

Fig. 10.

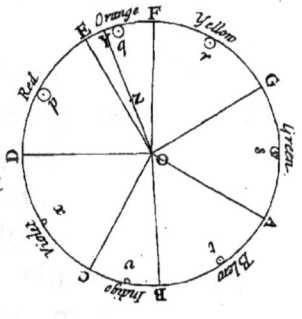

Fig. 11.

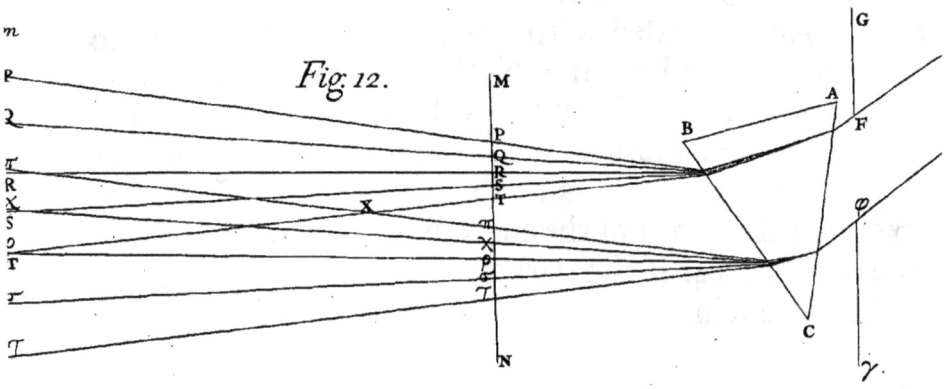

Fig. 12.

view the Colours made by a Prifm will find it fo in Nature.

These are the Colours on both fides the white when the Paper is held between the Prifm, and the point X where the Colours meet, and the interjacent white vanifhes. For if the Paper be held ftill farther off from the Prifm, the moft refrangible and leaft refrangible rays will be wanting in the middle of the Light, and the reft of the rays which are found there, will by mixture produce a fuller green than before. Alfo the yellow and blue will now become lefs compounded, and by confequence more intenfe than before. And this alfo agrees with experience.

And if one look through a Prifm upon a white Object encompaffed with blacknefs or darknefs, the reafon of the Colours arifing on the edges is much the fame, as will appear to one that fhall a little confider it. If a black Object be encompaffed with a white one, the Colours which appear through the Prifm are to be derived from the Light of the white one, fpreading into the Regions of the black, and therefore they appear in a contrary order to that, in which they appear when a white Object is furrounded with black. And the fame is to be underftood when an Object is viewed, whofe parts are fome of them lefs luminous than others. For in the Borders of the more and lefs luminous parts, Colours ought always by the fame Principles to arife from the excefs of the Light of the more luminous, and to be of the fame kind as if the darker parts were black, but yet to be more faint and dilute.

Q 2

What

What is said of Colours made by Prisms may be easily applied to Colours made by the Glasses of Telescopes, or Microscopes, or by the humours of the Eye. For if the Object-glass of a Telescope be thicker on one side than on the other, or if one half of the Glass, or one half of the Pupil of the Eye be covered with any opake substance: the Object-glass, or that part of it or of the Eye which is not covered, may be considered as a Wedge with crooked sides, and every Wedge of Glass, or other pellucid substance, has the effect of a Prism in refracting the Light which passes through it.

How the Colours in the 9th and 10th Experiments of the first Part arise from the different reflexibility of Light, is evident by what was there said. But it is observable in the 9th Experiment, that whilst the Sun's direct Light is yellow, the excess of the blue-making rays in the reflected Beam of Light MN, suffices only to bring that yellow to a pale white inclining to blue, and not to tinge it with a manifestly blue Colour. To obtain therefore a better blue, I used instead of the yellow Light of the Sun the white Light of the Clouds, by varying a little the Experiment as follows

EXPER. XVI.

Fig. 13.   Let HFG represent a Prism in the open Air, and S the Eye of the Spectator, viewing the Clouds by their Light coming into the Prism at the plane side FIGK, and reflected in it by its base HEIG, and thence going out through its plain side HEFK to the Eye. And when the Prism and Eye are conveniently placed, so that the Angles of incidence and reflexion at the base

may

may be about 40 degrees, the Spectator will see a Bow
M N of a blue Colour, running from one end of the
base to the other, with the concave side towards him,
and the part of the base I M N G beyond this Bow will
be brighter than the other part E M N H on the other
side of it.   This blue Colour M N being made by nothing else than by reflexion of a specular superficies,
seems so odd a Phænomenon, and so unaccountable for
by the vulgar Hypothesis of Philosophers, that I could
not but think it deserved to be taken notice of.   Now
for understanding the reason of it, suppose the plane
A B C to cut the plane sides and base of the Prism perpendicularly. From the Eye to the line B C, wherein that
plane cuts the base, draw the lines S p and S t, in the
Angles S p c 50 degr. $\frac{2}{5}$, and S t c 49 degr. $\frac{1}{28}$, and the
point p will be the limit beyond which none of the most
refrangible rays can pass through the base of the Prism,
and be refracted, whose incidence is such that they may
be reflected to the Eye; and the point t will be the like
limit for the least refrangible rays, that is, beyond
which none of them can pass through the base, whose
incidence is such that by reflexion they may come to the
Eye.   And the point r taken in the middle way between
p and t, will be the like limit for the meanly refrangible
rays.   And therefore all the refrangible rays which fall
upon the base beyond t, that is, between t and B, and
can come from thence to the Eye will be reflected thither: But on this side t, that is, between t and c, many
of these rays will be transmitted through the base.
And all the most refrangible rays which fall upon the
base beyond p, that is, between p and B, and can by
reflexion come from thence to the Eye, will be reflected
thither,

thither, but every where between t and c, many of
thele rays will get through the bafe and be refracted;
and the fame is to be underftood of the meanly refrangible rays on either fide of the point r.　Whence it follows, that the bafe of the Prifm muft every where between t and B, by a total reflexion of all forts of rays to
the Eye, look white and bright.　And every where
between *p* and C, by reafon of the tranfmiffion of many
rays of every fort, look more pale, obfcure and dark.
But at r, and in other places between p and t, where
all the more refrangible rays are reflected to the Eye,
and many of the lefs refrangible are tranfmitted, the
excefs of the moft refrangible in the reflected Light will
tinge that Light with their Colour, which is violet and
blue.　And this happens by taking the line C p r t B any
where between the ends of the Prifm H G and E I.

PROP. IX. PROB. IV.

*By the difcovered Properties of Light to explain the
Colours of the Rain-bow.*

This Bow never appears but where it Rains in the
Sun-fhine, and may be made artificially by fpouting up
Water which may break aloft, and fcatter into Drops,
and fall down like Rain.　For the Sun fhining upon thefe
Drops certainly caufes the Bow to appear to a Spectator ftanding in a due pofition to the Rain and Sun.　And
hence it is now agreed upon, that this Bow is made by
refraction of the Sun's Light in Drops of falling Rain.
This was underftood by fome of the Ancients, and of
late more fully difcovered and explained by the Famous
Antonius

Antonius de Dominis Archbishop of *Spilato*, in his Book
De Radiis Visus & Lucis, published by his Friend *Bar-*
tolus at *Venice*, in the Year 1611, and written above
twenty Years before. For he teaches there how the
interior Bow is made in round Drops of Rain by two
refractions of the Sun's Light, and one reflexion between them, and the exterior by two refractions and
two forts of reflexions between them in each Drop of
Water, and proves his Explications by Experiments
made with a Phial full of Water, and with Globes of Glafs
filled with Water, and placed in the Sun to make the
Colours of the two Bows appear in them. The fame
Explication *Des-Cartes* hath purfued in his Meteors,
and mended that of the exterior Bow. But whilft they
underftood not the true origin of Colours, it's neceffary
to purfue it here a little further. For underftanding
therefore how the Bow is made, let a Drop of Rain or
any other fpherical tranfparent Body be reprefented by
the Sphere B N F G, defcribed with the Center C, and *Fig.* 14.
Semi-diameter C N. And let A N be one of the Sun's
rays incident upon it at N, and thence refracted to F,
where let it either go out of the Sphere by refraction towards V, or be reflected to G; and at G let it either go
out by refraction to R, or be reflected to H; and at H
let it go out by refraction towards S, cutting the incident ray in Y; produce A N and R G, till they meet in
X, and upon A X and N F let fall the perpendiculars
C D and C E, and produce C D till it fall upon the circumference at L. Parallel to the incident ray A N draw
the Diameter B Q, and let the fine of incidence out of
Air into Water be to the fine of refraction as I to
R. Now if you fuppofe the point of incidence N to
move

move from the point B, continually till it come to L, the Arch Q F will firft increafe and then decreafe, and fo will the Angle A X R which the rays A N and G R contain ; and the Arch Q F and Angle A X R will be biggeft when N D is to C N as $\sqrt{II\text{-}RR}$ to $\sqrt{3}$ R R, in which cafe N E will be to N D as 2 R to I. Alfo the Angle A Y S which the rays A N and H S contain will firft decreafe, and then increafe and grow leaft when N D is to C N as $\sqrt{II\text{-}RR}$ to $\sqrt{8}$ R R, in which cafe N E will be to N D as 3 R to I. And fo the Angle which the next emergent ray (that is, the emergent ray after three reflexions) contains with the incident ray A N will come to its limit when N D is to C N as $\sqrt{II\text{-}RR}$ to $\sqrt{15}$ R R, in which cafe N E will be to N D as 4 R to I, and the Angle which the ray next after that emergent, that is, the ray emergent after four reflexions, con-tains with the incident will come to its limit, when N D is to C N, as $\sqrt{II\text{-}RR}$ to $\sqrt{24}$ R R, in which cafe N E will be to N D as 5 R to I ; and fo on infinitely, the numbers 3, 8, 15, 24, &c. being gathered by conti-nual addition of the terms of the arithmetical progreffion 3, 5, 7, 9, &c. The truth of all this Mathematicians will eafily examine.

Now it is to be obferved, that as when the Sun comes to his Tropicks, days increafe and decreafe but a very little for a great while together ; fo when by increafing the diftance C D, thefe Angles come to their limits, they vary their quantity but very little for fome time together, and therefore a far greater number of the rays which fall upon all the points N in the Quadrant B L, fhall emerge in the limits of thefe Angles, then in any other inclinations. And further it is

to

to be obferved, that the rays which differ in refrangibility will have different limits of their Angles of emergence, and by confequence according to their different
degrees of refrangibility emerge moft copioufly in different Angles, and being feparated from one another
appear each in their proper Colours. And what thofe
Angles are may be eafily gathered from the foregoing
Theorem by computation.

For in the leaft refrangible rays the fines I and R (as
was found above) are 108 and 81, and thence by
computation the greateft Angle A X R will be found
42 degrees and 2 minutes, and the leaft Angle A Y S,
50 degr. and 57 minutes. And in the moft refrangible
rays the fines I and R are 109 and 81, and thence by
computation the greateft Angle A X R will be found
40 degrees and 17 minutes, and the leaft Angle A Y S
54 degrees and 7 minutes.

Suppofe now that O is the Spectator's Eye, and OP a line *Fig.* 15.
drawn parallel to the Sun's rays, and let P O E, P O F,
P O G, P O H, be Angles of 40 degr. 17 min. 42 degr.
2 min. 50 degr. 57 min. and 54 degr. 7 min. refpectively,
and thefe Angles turned about their common fide O P,
fhall with their other fides O E, O F; O G, O H de-
defcribe the verges of two Rain-bows A F B E and
C H D G. For if E, F, G, H, be Drops placed any
where in the conical fuperficies defcribed by O E, O F,
O G, O H, and be illuminated by the Sun's rays S E,
S F, S G, S H; the Angle S E O being equal to the
Angle P O E or 40 degr. 17 min. fhall be the greateft
Angle in which the moft refrangible rays can after one
reflexion be refracted to the Eye, and therefore all the
Drops in the line O E fhall fend the moft refrangible

rays

rays moft copiouſly to the Eye, and thereby ſtrike the ſenſes with the deepeſt violet Colour in that region. And in like manner the Angle S F O being equal to the Angle P O F, or 42 deg. 2 min. ſhall be the greateſt in which the leaſt refrangible rays after one reflexion can emerge out of the Drops, and therefore thoſe rays ſhall come moſt copiouſly to the Eye from the Drops in the line O F, and ſtrike the ſenſes with the deepeſt red Colour in that region. And by the ſame argument, the rays which have intermediate degrees of refrangibility ſhall come moſt copiouſly from Drops between E and F, and ſtrike the ſenſes with the intermediate Colours in the order which their degrees of refrangibility require, that is, in the progreſs from E to F, or from the inſide of the Bow to the outſide in this order, violet, indico, blue, green, yellow, orange, red. But the violet, by the mixture of the white Light of the Clouds, will appear faint and incline to purple.

Again, the Angle S G O being equal to Angle P O G, or 50 gr. 51 min. ſhall be the leaſt Angle in which the leaſt refrangible rays can after two reflexions emerge out of the Drops, and therefore the leaſt refrangible rays ſhall come moſt copiouſly to the Eye from the Drops in the line O G, and ſtrike the ſenſe with the deepeſt red in that region. And the Angle S H O being equal to the Angle P O H or 54 gr. 7 min. ſhall be the leaſt Angle in which the moſt refrangible rays after two reflections can emerge out of the Drops, and therefore thoſe rays ſhall come moſt copiouſly to the Eye from the Drops in the line O H, and ſtrike the ſenſes with the deepeſt violet in that region. And by the ſame argument, the Drops in the regions between G and H ſhall ſtrike the ſenſe with

the

the intermediate Colours in the order which their degrees of refrangibility require, that is, in the progrefs from G to H, or from the infide of the Bow to the outfide in this order, red, orange, yellow, green, blue, indico, violet. And fince thefe four lines O E, O F, O G. O H, may be fituated any where in the above-mentioned conical fuperficies, what is faid of the Drops and Colours in thefe lines is to be underftood of the Drops and Colours every where in thofe fuperficies.

Thus fhall there be made two Bows of Colours, an interior and ftronger, by one reflexion in the Drops, and an exterior and fainter by two ; for the Light becomes fainter by every reflexion. And their Colours fhall ly in a contrary order to one another, the red of both Bows bordering upon the fpace G F which is between the Bows. The breadth of the interior Bow E O F meafured crofs the Colours fhall be 1 degr. 45 min. and the breadth of the exterior G O H fhall be 3 degr. 10 min. and the diftance between them G O F fhall be 8 gr. 55 min. the greateft Semi-diameter of the innermoft, that is, the Angle P O F being 42 gr. 2 min. and the leaft Semi-diameter of the outermoft P O G, being 50 gr. 57 min. Thefe are the meafures of the Bows, as they would be were the Sun but a point ; for by the breadth of his Body the breadth of the Bows will be increafed and their diftance decreafed by half a degree, and fo the breadth of the interior Iris will be 2 degr. 15 min. that of the exterior 3 degr. 40 min. their diftance 8 degr. 25 min. the greateft Semi-diameter of the interior Bow 42 degr. 17 min. and the leaft of the exterior 50 degr. 42 min. And fuch are the dimenfions of the Bows in the Heavens found to be very nearly,

R 2   when

when their Colours appear ftrong and perfect. For once, by fuch means as I then had, I meafured the greateft Semi-diameter of the interior Iris about 42 degrees, the breadth of the red, yellow and green in that Iris 63 or 64 minutes, befides the outmoft faint red obfcured by brightnefs of the Clouds, for which we may allow 3 or 4 minutes more. The breadth of the blue was about 40 minutes more befides the violet, which was fo much obfcured by the brightnefs of the Clouds, that I could not meafure its breadth. But fuppofing the breadth of the blue and violet together to equal that of the red, yellow and green together, the whole breadth of this Iris will be about $2\frac{1}{4}$ degrees as above. The leaft diftance between this Iris and the exterior Iris was about 8 degrees and 30 minutes. The exterior Iris was broader than the interior, but fo faint, efpecially on the blue fide, that I could not meafure its breadth diftinctly. At another time when both Bows appeared more diftinct, I meafured the breadth of the interior Iris 2 gr. 10', and the breadth of the red, yellow and green in the exterior Iris, was to the breadth of the fame Colours in the interior as 3 to 2.

This Explication of the Rain-bow is yet further confirmed by the known Experiment (made by *Antonius de Dominis* and *Des-Cartes*) of hanging up any where in the Sun-fhine a Glafs-Globe filled with Water, and viewing it in fuch a pofture that the rays which come from the Globe to the Eye may contain with the Sun's rays an Angle of either 42 or 50 degrees. For if the Angle be about 42 or 43 degrees, the Spectator (fuppofe at O) fhall fee a full red Colour in that fide of the Globe oppofed to the Sun as 'tis reprefented at F, and

if

if that Angle become lefs (fuppofe by deprefling the Globe to E) there will appear other Colours, yellow, green and blue fucceffively in the fame fide of the Globe. But if the Angle be made about 50 degrees (fuppofe by lifting up the Globe to G)there will appear a red Colour in that fide of the Globe towards the Sun, and if the Angle be made greater (fuppofe by lifting up the Globe to H) the red will turn fucceffively to the other Colours yellow, green and blue. The fame thing I have tried by letting a Globe reft, and raifing or deprefling the Eye, or otherwife moving it to make the Angle of a juft magnitude.

I have heard it reprefented, that if the Light of a Candle be refracted by a Prifm to the Eye ; when the blue Colour falls upon the Eye the Spectator fhall fee red in the Prifm, and when the red falls upon the Eye he fhall fee blue ; and if this were certain, the Colours of the Globe and Rain-bow ought to appear in a contrary order to what we find. But the Colours of the Candle being very faint, the miftake feems to arife from the difficulty of difcerning what Colours fall on the Eye. For, on the contrary, I have fometimes had occafion to obferve in the Sun's Light refracted by a Prifm, that the Spectator always fees that Colour in the Prifm which falls upon his Eye. And the fame I have found true alfo in Candle-Light. For when the Prifm is moved flowly from the line which is drawn directly from the Candle to the Eye, the red appears firft in the Prifm and then the blue, and therefore each of them is feen when it falls upon the Eye. For the red paffes over the Eye firft, and then the blue.

The

The Light which comes through Drops of Rain by two refractions without any reflexion, ought to appear ftrongeft at the diftance of about 26 degrees from the Sun, and to decay gradually both ways as the diftance from him increafes and decreafes. And the fame is to be underftood of Light tranfmitted through fpherical Hail-ftones. And if the Hail be a little flatted, as it often is, the Light tranfmitted may grow fo ftrong at a little lefs diftance than that of 26 degrees, as to form a Halo about the Sun or Moon; which Halo, as often as the Hail-ftones are duly figured may be coloured, and then it muft be red within by the leaft refrangible rays, and blue without by the moft refrangible ones, efpecially if the Hail-ftones have opake Globules of Snow in their center to intercept the Light within the Halo (as *Hugenius* has obferved) and make the infide thereof more diftinctly defined than it would otherwife be. For fuch Hail-ftones, though fpherical, by terminating the Light by the Snow, may make a Halo red within and colourlefs without, and darker in the red than without, as Halos ufe to be. For of thofe rays which pafs clofe by the Snow the rubriform will be leaft refracted, and fo come to the Eye in the directeft lines.

The Light which paffes through a Drop of rain after two refractions, and three or more reflexions, is fcarce ftrong enough to caufe a fenfible Bow ; but in thofe Cylinders of Ice by which *Hugenius* explains the *Parhelia*, it may perhaps be fenfible.

PROP.

PROP. X. PROB. V.

By the discovered Properties of Light to explain the permanent Colours of natural Bodies.

These Colours arise from hence, that some natural Bodies reflect some sorts of rays, others other sorts more copiously than the rest. Minium reflects the least refrangible or red-making rays most copiously, and thence appears red. Violets reflect the most refrangible, most copiously, and thence have their Colour, and so of other Bodies. Every Body reflects the rays of its own Colour more copiously than the rest, and from their excess and predominance in the reflected Light has its Colour.

EXPER. XVII.

For if the homogeneal Lights obtained by the solution of the Problem proposed in the 4th Proposition of the first Book you place Bodies of several Colours, you will find, as I have done, that every Body looks most splendid and luminous in the Light of its own Colour. Cinnaber in the homogeneal red Light is most resplendent, in the green Light it is manifestly less resplendent, and in the blue Light still less. Indico in the violet blue Light is most resplendent, and its splendor is gradually diminished as it is removed thence by degrees through the green and yellow Light to the red. By a Leek the green Light, and next that the blue and yellow which compound green, are more strongly reflected

flected than the other Colours red and violet, and so of the rest. But to make these Experiments the more manifest, such Bodies ought to be chosen as have the fullest and most vivid Colours, and two of those Bodies are to be compared together. Thus, for instance, if Cinnaber and *ultra* marine blue, or some other full blue be held together in the homogeneal Light, they will both appear red, but the Cinnaber will appear of a strongly luminous and resplendent red, and the *ultra* marine blue of a faint obscure and dark red; and if they be held together in the blue homogeneal Light they will both appear blue, but the *ultra* marine will appear of a strongly luminous and resplendent blue, and the Cinnaber of a faint and dark blue. Which puts it out of dispute, that the Cinnaber reflects the red Light much more copiously than the *ultra* marine doth, and the *ultra* marine reflects the blue Light much more copiously than the Cinnaber doth. The same Experiment may be tryed succesfully with red Lead and Indico, or with any other two coloured Bodies, if due allowance be made for the different strength or weakness of their Colour and Light.

And as the reason of the Colours of natural Bodies is evident by these Experimenrs, so it is further confirmed and put past dispute by the two first Experiments of the first Book, whereby 'twas proved in such Bodies that the reflected Light which differ in Colours do differ also in degrees of refrangibility. For thence it's certain, that some Bodies reflect the more refrangible, others the less refrangible rays more copiously.

And

And that this is not only a true reason of these Colours, but even the only reason may appear further from this consideration, that the Colour of homogeneal Light cannot be changed by the reflexion of natural Bodies.

For if Bodies by reflexion cannot in the least change the Colour of any one sort of rays, they cannot appear coloured by any other means than by reflecting those which either are of their own Colour, or which by mixture must produce it.

But in trying Experiments of this kind care must be had that the Light be sufficiently homogeneal. For if Bodies be illuminated by the ordinary prismatick Colours, they will appear neither of their own day-light Colours, nor of the Colour of the Light cast on them, but of some middle Colour between both, as I have found by Experience. Thus red Lead (for instance) illuminated with the ordinary prismatick green will not appear either red or green, but orange or yellow, or between yellow and green accordingly, as the green Light by which 'tis illuminated is more or less compounded. For because red Lead appears red when illuminated with white Light, wherein all sorts of rays are equally mixed, and in the green Light all sorts of rays are not equally mixed, the excess of the yellow-making, green-making and blue-making rays in the incident green Light, will cause those rays to abound so much in the reflected Light as to draw the Colour from red towards their Colour. And because the red Lead reflects the red-making rays most copiously in proportion to their number, and next after them the orange-making and yellow-making rays; these rays in

S   the

the reflected Light will be more in proportion to the Light than they were in the incident green Light, and thereby will draw the reflected Light from green towards their Colour. And therefore the red Lead will appear neither red nor green, but of a Colour between both.

In transparently coloured Liquors 'tis observable, that their Colour uses to vary with their thickness. Thus, for instance, a red Liquor in a conical Glass held between the Light and the Eye, looks of a pale and dilute yellow at the bottom where 'tis thin, and a little higher where 'tis thicker grows orange, and where 'tis still thicker becomes red, and where 'tis thickest the red is deepest and darkest. For it is to be conceived that such a Liquor stops the indico-making and violet-making rays most easily, the blue-making rays more difficultly, the green-making rays still more difficultly, and the red-making most difficultly : And that if the thickness of the Liquor be only so much as suffices to stop a competent number of the violet-making and indico-making rays, without diminishing much the number of the rest, the rest must (by Prop. 6. *Lib.* 2.) compound a pale yellow. But if the Liquor be so much thicker as to stop also a great number of the blue-making rays, and some of the green-making, the rest must compound an orange ; and where it is so thick as to stop also a great number of the green-making and a considerable number of the yellow-making, the rest must begin to compound a red, and this red must grow deeper and darker as the yellow making and orange-making rays are more and more stopt by increasing the thickness of the Liquor, so that few rays besides the red-making can get through.

Of

Of this kind is an Experiment lately related to me by Mr. *Halley*, who, in diving deep into the Sea, found in a clear Sun-fhine day, that when he was funk many Fathoms deep into the Water, the upper part of his Hand in which the Sun fhone directly through the Water looked of a red Colour, and the under part of his Hand illuminated by Light reflected from the Water below looked green. For thence it may be gathered, that the Sea-water reflects back the violet and blue-making rays moft eafily, and lets the red-making rays pafs moft freely and copioufly to great depths. For thereby the Sun's direct Light at all great depths, by reafon of the predominating red-making rays, muft appear red ; and the greater the depth is, the fuller and intenfer muft that red be. And at fuch depths as the violet-making rays fcarce penetrate unto, the blue-making, green-making and yellow-making rays being reflected from below more copioufly than the red-making ones, muft compound a green.

Now if there be two Liquors of full Colours, fuppofe a red and a blue, and both of them fo thick as fuffices to make their Colours fufficiently full ; though either Liquor be fufficiently tranfparent apart, yet will you not be able to fee through both together. For if only the red-making rays pafs through one Liquor, and only the blue-making through the other, no rays can pafs through both. This Mr. *Hook* tried cafually with Glafs-wedges filled with red and blue Liquors, and was furprized at the unexpected event, the reafon of it being then unknown ; which makes me truft the more to his Experiment, though I have not tryed it my felf. But he that would repeat it, muft take care the Liquors be of very good and full Colours.

Now

Now whilft Bodies become coloured by reflecting or
tranfmitting this or that fort of rays more copioufly than
the reft, it is to be conceived that they ftop and ftifle in
themfelves the rays which they do not reflect or tranfmit.
For if Gold be foliated and held between your Eye and
the Light, the Light looks blue, and therefore maffy Gold
lets into its Body the blue-making rays to be reflected
to and fro within it till they be ftopt and ftifled, whilft
it reflects the yellow-making outwards, and thereby
looks yellow. And much after the fame manner that
Leaf-gold is yellow by reflected, and blue by tranfmitted Light, and maffy Gold is yellow in all pofitions of
the Eye; there are fome Liquors as the tincture of
Lignum Nephriticum, and fome forts of Glafs which
tranfmit one fort of Light moft copioufly, and reflect
another fort, and thereby look of feveral Colours, according to the pofition of the Eye to the Light. But if
thefe Liquors or Glaffes were fo thick and maffy that
no Light could get through them, I queftion not but
that they would like all other opake Bodies appear of
one and the fame Colour in all pofitions of the Eye,
though this I cannot yet affirm by experience. For all
coloured Bodies, fo far as my Obfervation reaches, may
be feen through if made fufficiently thin, and therefore
are in fome meafure tranfparent, and differ only in degrees of tranfparency from tinged tranfparent Liquors;
thefe Liquors, as well as thofe Bodies, by a fufficient
thicknefs becoming opake. A tranfparent Body which
looks of any Colour by tranfmitted Light, may alfo
look of the fame Colour by reflected Light, the Light
of that Colour being reflected by the further furface of
the Body, or by the Air beyond it. And then the reflected Colour will be diminifhed, and perhaps ceafe, by

making

making the Body very thick, and pitching it on the back-side to diminish the reflexion of its further surface, so that the Light reflected from the tinging particles may predominate. In such cases, the Colour of the reflected Light will be apt to vary from that of the Light transmitted. But whence it is that tinged Bodies and Liquors reflect some sort of rays, and intromit or transmit other sorts, shall be said in the next Book. In this Proposition I content my self to have put it past dispute, that Bodies have such Properties, and thence appear coloured.

PROP. XI. PROB. VI.

By mixing coloured Lights to compound a Beam of Light of the same Colour and Nature with a Beam of the Sun's direct Light, and therein to experience the truth of the foregoing Propositions.

Let A B C a b c represent a Prism by which the Sun's *Fig.* 16. Light let into a dark Chamber through the Hole F, may be refracted towards the Lens M N, and paint upon it at p, q, r, s and t, the usual Colours violet, blue, green, yellow and red, and let the diverging rays by the refraction of this Lens converge again towards X, and there, by the mixture of all those their Colours, compound a white according to what was shewn above. Then let another Prism D E G d e g, parallel to the former, be placed at X, to refract that white Light upwards towards Y. Let the refracting Angles of the Prisms, and their distances from the Lens be equal, so that the rays which converged from the Lens towards X, and without refraction, would there have crossed and diverged again, may by the refraction of the second Prism be

reduced

reduced into Parallelifm and diverge no more. For then thofe rays will recompofe a Beam of white Light X Y. If the refracting Angle of either Prifm be the bigger, that Prifm muft be fo much the nearer to the Lens. You will know when the Prifms and the Lens are well fet together by obferving if the Beam of Light X Y which comes out of the fecond Prifm be perfectly white to the very edges of the Light, and at all diftances from the Prifm continue perfectly and totally white like a Beam of the Sun's Light. For till this happens, the pofition of the Prifms and Lens to one another muft be corrected, and then if by the help of a long Beam of Wood, as is reprefented in the Figure, or by a Tube, or fome other fuch inftrument made for that purpofe, they be made faft in that fituation, you may try all the fame Experiments in this compounded Beam of Light X Y, which in the foregoing Experiments have been made in the Sun's direct Light. For this compounded Beam of Light has the fame appearance, and is endowed with all the fame Properties with a direct Beam of the Sun's Light, fo far as my Obfervation reaches. And in trying Experiments in this Beam you may by ftopping any of the Colours p, q, r, s and t, at the Lens, fee how the Colours produced in the Experiments are no other than thofe which the rays had at the Lens before they entered the compofition of this Beam : And by confequence that they arife not from any new modifications of the Light by refractions and reflexions, but from the various feparations and mixtures of the rays originally endowed with their colour-making qualities.

So, for inftance, having with a Lens 4¼ Inches broad, and two Prifms on either Hand 6¼ Feet diftant from the Lens, made fuch a Beam of compounded Light : to
examin

examin the reafon of the Colours made by Prifms, I refracted this compounded Beam of Light X Y with another Prifm H I K k h, and thereby caft the ufual prifmatick Colours P Q R S T upon the Paper L V placed behind. And then by ftopping any of the Colours p, q, r, s, t, at the Lens, I found that the fame Colour would vanifh at the Paper. So if the purple P was ftopped at the Lens, the purple P upon the Paper would vanifh, and the reft of the Colours would remain unaltered, unlefs perhaps the blue, fo far as fome purple latent in it at the Lens might be feparated from it by the following refractions. And fo by intercepting the green upon the Lens, the green R upon the Paper would vanifh, and fo of the reft ; which plainly fhews, that as the white Beam of Light X Y was compounded of feve Lights varioufly coloured at the Lens, fo the Colours which afterwards emerge out of it by new refractions are no other than thofe of which its whitenefs was compounded. The refraction of the Prifm H I K k h generates the Colours P Q R S T upon the Paper, not by changing the colorific qualities of the rays, but by feparating the rays which had the very fame colorific qualities before they entered the compofition of the refracted Beam white of Light X Y. For otherwife the rays which were of one Colour at the Lens might be of another upon the Paper, contrary to what we find.

So again, to examin the reafon of the Colours of natural Bodies, I placed fuch Bodies in the Beam of Light X Y, and found that they all appeared there of thofe their own Colours which they have in Day-light, and that thofe Colours depend upon the rays which had the fame Colours at the Lens before they entred the compofition

fition of that Beam. Thus, for inftance, Cinnaber illuminated by this Beam appears of the fame red Colour as in
Day-light ; and if at the Lens you intercept the green-
making and blue-making rays, its rednefs will become
more full and lively : But if you there intercept the red-
making rays, it will not any longer appear red, but become yellow or green, or of fome other Colour, according to the forts of rays which you do not intercept.
So Gold in this Light X Y appears of the fame yellow
Colour as in Day-light, but by intercepting at the Lens a
due quantity of the yellow-making rays it will appear
white like Silver (as I have tryed) which fhews that its
yellownefs arifes from the excefs of the intercepted rays
tinging that whitenefs with their Colour when they are
let pafs. So the infufion of *Lignum Nephriticum* (as I
have alfo tryed) when held in this Beam of Light X Y,
looks blue by the reflected part of the Light, and yellow
by the tranfmitted part of it, as when 'tis viewed in Day-
light, but if you intercept the blue at the Lens the infufion will lofe its reflected blue Colour, whilft its tranfmitted red remains perfect and by the lofs of fome blue-
making rays wherewith it was allayed becomes more intenfe and full. And, on the contrary, if the red and orange-
making rays be intercepted at Lens, the infufion will
lofe its tranfmitted red, whilft its blue will remain and
become more full and perfect. Which fhews, that the infufion does not tinge the rays with blue and yellow, but
only tranfmit thofe moft copioufly which were red-making before, and reflects thofe moft copioufly which were
blue-making before. And after the fame manner may the
reafons of other Phænomena be examined, by trying
them in this artificial Beam of Light X Y.

THE

Fig. 13.

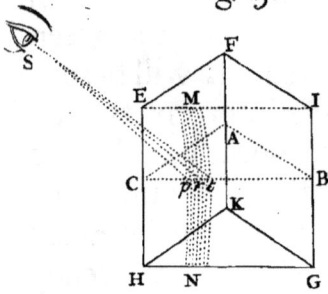

Fig. 14.

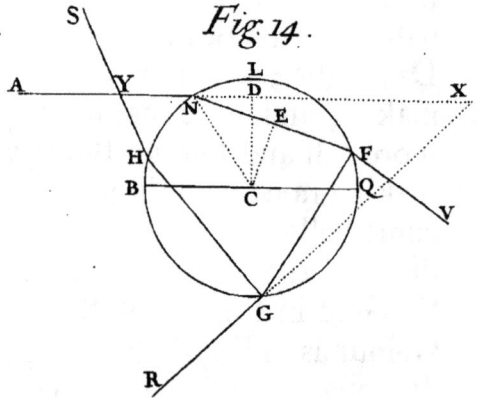

Fig. 15.

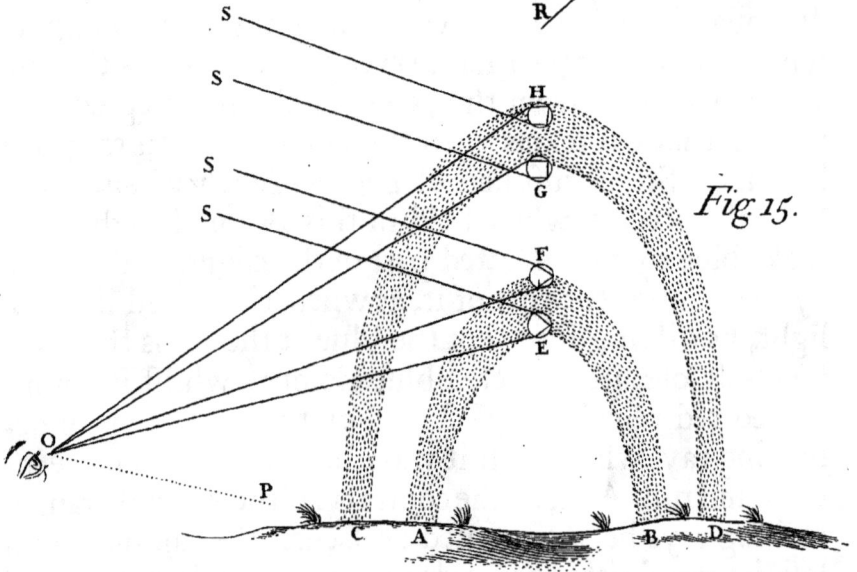

Fig. 16.

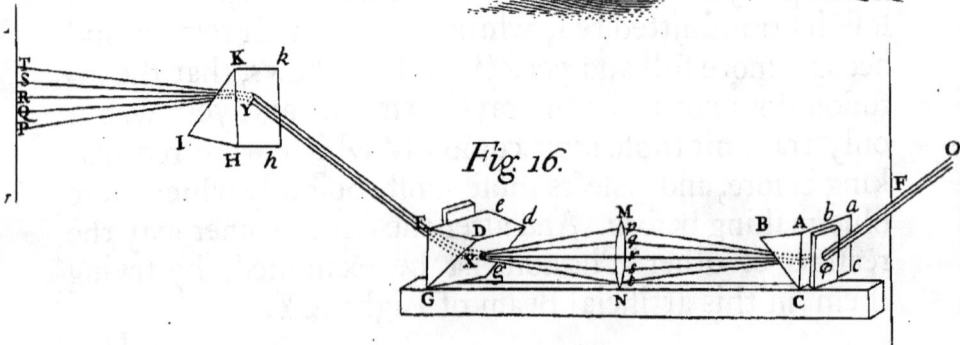

THE

SECOND BOOK

O F

OPTICKS.

PART I.

Obſervations concerning the Reflexions, Refractions, and Colours of thin tranſparent Bodies.

IT has been obſerved by others that tranſparent Subſtances, as Glaſs, Water, Air, &c. when made very thin by being blown into Bubbles, or otherwiſe formed into Plates, do exhibit various Colours according to their various thinneſs, although at a greater thickneſs they appear very clear and colourleſs. In the former Book I forbore to treat of theſe Colours, becauſe they ſeemed of a more difficult conſideration, and were not neceſſary for eſtabliſhing the Properties of Light there diſcourſed of. But becauſe they may conduce to further diſcoveries for completing the Theory of Light, eſpecially as to the conſtitution of the parts of natural Bodies, on which their Colours or Tranſparency depend ; I have here ſet down an account of them. To render this Diſcourſe ſhort and diſtinct, I have firſt deſcribed the principal of my

A a   Obſer-

Obfervations, and then confidered and made ufe of them. The Obfervations are thefe.

O B S. I.

Compreffing two Prifms hard together that their Sides (which by chance were a very little convex)might fomewhere touch one another: I found the place in which they touched to become abfolutely tranfparent, as if they had there been one continued piece of Glafs. For when the Light fell fo obliquely on the Air, which in other places was between them,as to be all reflected ; it feemed in that place of contact to be wholly tranf-mitted, infomuch that when looked upon, it appeared like a black or dark Spot, by reafon that little or no fenfible Light was reflected from thence, as from other places; and when looked through it feemed (as it were) a hole in that Air which was formed into a thin Plate, by being compreffed between the Glaffes. And through this hole Objects that were beyond might be feen di-ftinctly, which could not at all be feen through other parts of the Glaffes where the Air was interjacent. Al-though the Glaffes were a little convex, yet this tranf-parent Spot was of a confiderable breadth,which breadth feemed principally to proceed from the yielding inwards of the parts of the Glaffes, by reafon of their mutual preffure. For by preffing them very hard together it would become much broader than otherwife.

O B S.

O B S. II.

When the Plate of Air, by turning the Prisms about their common Axis, became so little inclined to the incident Rays, that some of them began to be transmitted, there arose in it many slender Arcs of Colours which at first were shaped almost like the Conchoid, as you see them delineated in the first Figure. And *Fig.* 1. by continuing the motion of the Prisms, these Arcs increased and bended more and more about the said transparent Spot, till they were completed into Circles or Rings incompassing it, and afterwards continually grew more and more contracted.

These Arcs at their first appearance were of a violet and blue Colour, and between them were white Arcs of Circles, which presently by continuing the motion of the Prisms became a little tinged in their inward Limbs with red and yellow, and to their outward Limbs the blue was adjacent. So that the order of these Colours from the central dark Spot, was at that time white, blue, violet ; black ; red, orange, yellow, white, blue, violet, &c. But the yellow and red were much fainter than the blue and violet.

The motion of the Prisms about their Axis being continued, these Colours contracted more and more, shrinking towards the whiteness on either side of it, until they totally vanished into it. And then the Circles in those parts appeared black and white, without any other Colours intermixed. But by further moving the Prisms about, the Colours again emerged out of the whiteness, the violet and blue as its inward Limb, and at its out-

ward

ward Limb the red and yellow. So that now their order from the central Spot was white, yellow, red ; black ; violet, blue, white, yellow, red, &c. contrary to what it was before.

O B S. III.

When the Rings or fome parts of them appeared only black and white, they were very diftinct and well defined, and the backnefs feemed as intenfe as that of the central Spot. Alfo in the borders of the Rings, where the Colours began to emerge out of the whitenefs, they were pretty diftinct, which made them vifible to a very great Multitude. I have fometimes numbred above thirty Succeffions (reckoning every black and white Ring for one Succeffion) and feen more of them, which by reafon of their fmalnefs I could not number. But in other Pofitions of the Prifms, at which the Rings appeared of many Colours, I could not diftinguifh above eight or nine of them, and the exterior of thofe were very confufed and dilute.

In thefe two Obfervations to fee the Rings diftinct, and without any other Colour than black and white, I found it neceffary to hold my Eye at a good diftance from them. For by approaching nearer, although in the fame inclination of my Eye to the plane of the Rings, there emerged a blueifh Colour out of the white, which by dilating it felf more and more into the black rendred the Circles lefs diftinct, and left the white a little tinged with red and yellow. I found alfo by looking through a flit or oblong hole , which was narrower than the Pupil of my Eye, and held clofe to

it

it parallel to the Prifms, I could fee the Circles much diftincter and vifible to a far greater number than otherwife.

O B S. IV.

To obferve more nicely by the order of the Colours which arofe out of the white Circles as the Rays became lefs and lefs inclined to the plate of Air; I took two Object Glaffes, the one a Plano-convex for a fourteen-foot Telefcope, and the other a large double convex for one of about fifty-foot; and upon this, laying the other with its its plane-fide downwards, I preffed them flowly together, to make the Colours fucceffively emerge in the middle of the Circles, and then flowly lifted the upper Glafs from the lower to make them fucceffively vanifh again in the fame place. The Colour, which by preffing the Glaffes together emerged laft in the middle of the other Colours, would upon its firft appearance look like a Circle of a Colour almoft uniform from the circumference to the center, and by compreffing the Glaffes ftill more, grow continually broader until a new Colour emerged in its center, and thereby it became a Ring encompaffing that new Colour. And by compreffing the Glaffes ftill more, the Diameter of this Ring would encreafe, and the breadth of its Orbit or Perimeter decreafe until another new Colour emerged in the center of the laft : And fo on until a third, a fourth, a fifth, and other following new Colours fucceffively emerged there, and became Rings encompaffing the innermoft Colour, the laft of which was the black Spot. And, on the contrary, by

lifting

lifting up the upper Glafs from the lower, the diameter of the Rings would decreafe, and the breadth of their Orbit encreafe, until their Colours reached fucceffively to the center ; and then they being of a confiderable breadth, I could more eafily difcern and diftinguifh their Species than before. And by this means I obferved their Succeffion and Quantity to be as followeth.

Next, to the pellucid central Spot made by the contact of the Glaffes fucceeded blue, white, yellow, and red, the blue was fo little in quantity that I could not difcern it in the circles made by the Prifms, nor could I well diftinguifh any violet in it, but the yellow and red were pretty copious, and feemed about as much in extent as the white, and four or five times more than the blue. The next Circuit in order of Colours immediately encompaffing thefe were violet, blue, green, yellow, and red, and thefe were all of them copious and vivid, excepting the green, which was very little in quantity, and feemed much more faint and dilute than the other Colours. Of the other four, the violet was the leaft in extent, and the blue lefs than the yellow or red. The third Circuit or Order was purple, blue, green, yellow, and red ; in which the purple feemed more reddifh than the violet in the former Circuit, and the green was much more confpicuous, being as brifque and copious as any of the other Colours, except the yellow ; but the red began to be a little faded, inclining very much to purple. After this fucceeded the fourth Circuit of green and red. The green was very copious and lively, inclining on the one fide to blue, and on the other fide to yellow. But in

this

this fourth Circuit there was neither violet, blue, nor yellow, and the red was very imperfect and dirty Also the succeeding Colours became more and more imperfect and dilute, till after three or four Revolutions they ended in perfect whiteness. Their Form, when the Glasses were most compressed so as to make the black Spot appear in the Center, is delineated in the Second Figure; where $a, b, c, d, e : f, g, h, i, k : l, m, n, o, p : q, r :$ Fig. 2. $s, t : v, x : y$ denote the Colours reck'ned in order from the center, black, blue, white, yellow, red : violet, blue, green, yellow, red : purple, blue, green, yellow, red : green, red : greenish blue, red : greenish blue, pale red : greenish blue, reddish white.

O B S. V.

To determine the interval of the Glasses, or thickness of the interjacent Air, by which each Colour was produced, I measured the Diameters of the first six Rings at the most lucid part of their Orbits, and squaring them, I found their Squares to be in the Arithmetical Progression of the odd Numbers, 1.3.5.7.9.11. And since one of these Glasses was Plain, and the other Spherical, their Intervals at those Rings must be in the same Progression. I measured also the Diameters of the dark or faint Rings between the more lucid Colours, and found their Squares to be in the Arithmetical Progression of the even Numbers, 2.4.6.8.10.12. And it being very nice and difficult to take these measures exactly; I repeated them at divers times at divers parts of the Glasses, that by their Agreement I might be confirmed in them. And the same Method I used in

deter-

determining some others of the following Observations.

O B S. V I.

The Diameter of the sixth Ring at the most lucid part of its Orbit was $\frac{58}{100}$ parts of an Inch, and the Diameter of the Sphere on which the double convex Object-Glass was ground was about 102 Feet, and hence I gathered the thickness of the Air or Aereal Interval of the Glasses at that Ring. But some time after, suspecting that in making this Observation I had not determined the Diameter of the Sphere with sufficient accurateness, and being uncertain whether the Plano-convex Glass was truly plain, and not something concave or convex on that side which I accounted plain ; and whether I had not pressed the Glasses together, as I often did, to make them touch. (for by pressing such Glasses together their parts easily yield inwards, and the Rings thereby become sensibly broader than they would be, did the Glasses keep their Figures.) I repeated the Experiment, and found the Diameter of the sixth lucid Ring about $\frac{55}{100}$ parts of an Inch. I repeated the Experiment also with such an Object-Glass of another Telescope as I had at hand. This was a double convex ground on both sides to one and the same Sphere, and its Focus was distant from it $83\frac{1}{2}$ Inches. And thence, if the Sines of incidence and refraction of the bright yellow Light be assumed in proportion as 11 to 17, the Diameter of the Sphere to which the Glass was figured will by computation be found 182 Inches. This Glass I laid upon a flat one, so that the

black

black Spot appeared in the middle of the Rings of Colours without any other preſſure than that of the weight of the Glaſs. And now meaſuring the Diameter of the fifth dark Circle as accurately as I could, I found it the fifth part of an Inch preciſely. This meaſure was taken with the points of a pair of Compaſſes on the upper ſurface on the upper Glaſs, and my Eye was about eight or nine Inches diſtance from the Glaſs, almoſt perpendicularly over it, and the Glaſs was $\frac{1}{6}$ of an Inch thick, and thence it is eaſy to collect that the true Diameter of the Ring between the Glaſſes was greater than its meaſured Diameter above the Glaſſes in the proportion of 80 to 79 or thereabouts, and by conſequence equal to $\frac{16}{79}$ parts of an Inch, and its true Semi-diameter equal to $\frac{8}{79}$ parts. Now as the Diameter of the Sphere (182 Inches) is to the Semi-diameter of this fifth dark Ring ($\frac{8}{79}$ parts of an Inch) ſo is this Semi-diameter to the thickneſs of the Air at this fifth dark Ring ; which is therefore $\frac{32}{567931}$ or $\frac{10}{1774534}$ parts of an Inch, and the fifth part thereof ; viz. the $\frac{1}{88717}$th part of an Inch, is the thickneſs of the Air at the firſt of theſe dark Rings.

The ſame Experiment I repeated with another double convex Object-glaſs ground on both ſides to one and the ſame Sphere. Its Focus was diſtant from it 168$\frac{1}{2}$ Inches, and therefore the Diameter of that Sphere was 184 Inches. This Glaſs being laid upon the ſame plain Glaſs, the Diameter of the fifth of the dark Rings, when the black Spot in their center appeared plainly without preſſing the Glaſſes, was by the meaſure of the Compaſſes upon the upper Glaſs $\frac{121}{600}$ parts of an Inch, and by conſequence between the Glaſſes it was $\frac{1222}{6000}$. For the upper Glaſs was $\frac{1}{8}$ of an Inch thick,

B b   and

and my Eye was diftant from it 8 Inches. And a third proportional to half this from the Diameter of the Sphere is $\frac{5}{88850}$ parts of an Inch. This is therefore the thicknefs of the Air at this Ring, and a fifth part there-of, *viz.* the $\frac{1}{88850}$th part of an Inch is the thicknefs there-of at the firft of the Rings as above.

I tryed the fame thing by laying thefe Object-Glaffes upon flat pieces of a broken Looking-glafs, and found the fame meafures of the Rings : Which makes me rely upon them till they can be determined more ac-curately by Glaffes ground to larger Spheres, though in fuch Glaffes greater care muft be taken of a true plain.

Thefe Dimenfions were taken when my Eye was placed almoft perpendicularly over the Glaffes, being about an Inch, or an Inch and a quarter, diftant from the incident rays, and eight Inches diftant from the Glafs ; fo that the rays were inclined to the Glafs in an Angle of about 4 degrees. Whence by the following Obfervation you will underftand, that had the rays been perpendicular to the Glaffes, the thicknefs of the Air at thefe Rings would have been lefs in the propor-tion of the Radius to the fecant of 4 degrees, that is of 10000. Let the thickneffes found be therefore dimi-nifhed in this proportion, and they will become $\frac{1}{88940}$ and $\frac{1}{89063}$, or (to ufe the neareft round number) the $\frac{1}{89000}$th part of an Inch. This is the thicknefs of the Air at the darkeft part of the firft dark Ring made by perpendi-cular rays, and half this thicknefs multiplied by the progreffion, 1, 3, 5, 7, 9, 11, &c. gives the thickneffes of the Air at the moft luminous parts of all the brighteft Rings, *viz.* $\frac{1}{178000}$, $\frac{3}{178000}$, $\frac{5}{178000}$, $\frac{7}{178000}$, &c. their arithmetical means

means $\frac{2}{178000}$, $\frac{4}{178000}$, $\frac{6}{178000}$, &c. being its thicknesses at the darkest parts of all the dark ones.

O B S. VII.

The Rings were least when my Eye was placed perpendicularly over the Glasses in the Axis of the Rings: And when I viewed them obliquely they became bigger, continually swelling as I removed my Eye further from the Axis. And partly by measuring the Diameter of the same Circle at several obliquities of my Eye, partly by other means, as also by making use of the two Prisms for very great obliquities. I found its Diameter, and consequently the thickness of the Air at its perimeter in all those obliquities to be very nearly in the proportions expressed in this Table.

Angle of Incidence on the Air.		Angle of Refraction into the Air.		Diameter of the Ring.	Thickness of the Air.
deg.	min.				
00	00	00	00	10	10
06	26	10	00	$10\frac{1}{13}$	$10\frac{2}{13}$
12	45	20	00	$10\frac{2}{3}$	$10\frac{2}{3}$
18	49	30	00	$10\frac{3}{4}$	$11\frac{1}{2}$
24	30	40	00	$11\frac{2}{5}$	13
29	37	50	00	$12\frac{1}{2}$	$15\frac{1}{2}$
33	58	60	00	14	20
35	47	65	00	$15\frac{1}{4}$	$23\frac{1}{4}$
37	19	70	00	$16\frac{4}{5}$	$28\frac{1}{4}$
38	33	75	00	$19\frac{1}{4}$	37
39	27	80	00	$22\frac{6}{7}$	$52\frac{1}{4}$
40	00	85	00	29	$84\frac{1}{10}$
40	11	90	00	35	$122\frac{1}{2}$

In

In the two firſt Columns are expreſſed the obliquities of the incident and emergent rays to the plate of the Air, that is, their angles of incidence and refraction. In the third Column the Diameter of any coloured Ring at thoſe obliquities is expreſſed in parts, of which ten conſtitute that Diameter when the rays are perpendicular. And in the fourth Column the thickneſs of the Air at the circumference of that Ring is expreſſed in parts of which alſo ten conſtitute that thickneſs when the rays are perpendicular.

And from theſe meaſures I ſeem to gather this Rule: That the thickneſs of the Air is proportional to the ſecant of an angle, whoſe Sine is a certain mean proportional between the Sines of incidence and refraction. And that mean proportional, ſo far as by theſe meaſures I can determine it, is the firſt of an hundred and ſix arithmetical mean proportionals between thoſe Sines counted from the Sine of refraction when the refraction is made out of the Glaſs into the plate of Air, or from the Sine of incidence when the refraction is made out of the plate of Air into the Glaſs.

O B S. VIII.

The dark Spot in the middle of the Rings increaſed alſo by the obliquation of the Eye, although almoſt inſenſibly. But if inſtead of the Object-Glaſſes the Priſms were made uſe of, its increaſe was more manifeſt when viewed ſo obliquely that no Colours appeared about it. It was leaſt when the rays were incident moſt obliquely on the interjacent Air, and as the obliquity decreaſed it increaſed more and more until the coloured Rings appeared,

peared, and then decreafed again, but not fo much as
it increafed before. And hence it is evident, that the
tranfparency was not only at the abfolute contact of the
Glaffes, but alfo where they had fome little interval.
I have fometimes obferved the Diameter of that Spot to
be between half and two fifth parts of the Diameter of
the exterior circumference of the red in the firft circuit or revolution of Colours when viewed almoft perpendicularly ; whereas when viewed obliquely it hath
wholly vanifhed and become opake and white like the
other parts of the Glafs ; whence it may be collected
that the Glaffes did then fcarcely, or not at all, touch
one another, and that their interval at the perimeter
of that Spot when viewed perpendicularly was about a
fifth or fixth part of their interval at the circumference
of the faid red.

O B S. IX.

By looking through the two contiguous Object-
Glaffes, I found that the interjacent Air exhibited Rings
of Colours, as well by tranfmitting Light as by reflecting it. The central Spot was now white, and from it
the order of the Colours were yellowifh red ; black ;
violet, blue, white, yellow, red ; violet, blue, green,
yellow, red, &c. But thefe Colours were very faint
and dilute unlefs when the Light was trajected very
obliquely through the Glaffes : For by that means they
became pretty vivid. Only the firft yellowifh red, like
the blue in the fourth Obfervation, was fo little and
faint as fcarcely to be difcerned. Comparing the coloured Rings made by reflexion, with thefe made by
tranf-

transmission of the Light ; I found that white was opposite to black, red to blue, yellow to violet, and green
to a compound of red and violet.   That is, those parts
of the Glass were black when looked through, which
when looked upon appeared white, and on the contrary.   And so those which in one case exhibited blue,
did in the other case exhibit red.   And the like of the
Fig. 3. other Colours.   The manner you have represented in
the third Figure, where A B, C D, are the surfaces of
the Glasses contiguous at E, and the black lines between them are their distances in arithmetical progression, and the Colours written above are seen by reflected Light, and those below by Light transmitted.

O B S.   X.

Wetting the Object-Glasses a little at their edges,
the water crept in flowly between them, and the Circles thereby became less and the Colours more faint :
Infomuch that as the water crept along one half of
them at which it first arrived would appear broken off
from the other half, and contracted into a less room.
By measuring them I found the proportions of their
Diameters to the Diameters of the like Circles made by
Air to be about seven to eight, and consequently the intervals of the Glasses at like Circles, caused by those
two mediums Water and Air, are as about three to four.
Perhaps it may be a general Rule, That if any other
medium more or less dense than water be compressed
between the Glasses, their intervals at the Rings caused
thereby will be to their intervals caused by interjacent
Air,

Air, as the Sines are which meafure the refraction made out of that medium into Air.

O B S. XI.

When the water was between the Glaffes, if I pref-fed the upper Glafs varioufly at its edges to make the Rings move nimbly from one place to another, a little white Spot would immediately follow the center of them, which upon creeping in of the ambient water into that place would prefently vanifh. Its appearance was fuch as interjacent Air would have caufed, and it exhibited the fame Colours. But it was not Air, for where any bubbles of Air were in the water they would not vanifh. The reflexion muft have rather been caufed by a fubtiler medium, which could recede through the Glaffes at the creeping in of the water.

O B S. XII.

Thefe Obfervations were made in the open Air. But further to examin the effects of coloured Light falling on the Glaffes, I darkened the Room, and viewed them by reflexion of the Colours of a Prifm caft on a Sheet of white Paper, my Eye being fo placed that I could fee the coloured Paper by reflexion in the Glaffes, as in a Looking-glafs. And by this means the Rings be-came diftincter and vifible to a far greater number than in the open Air. I have fometimes feen more than twenty of them, whereas in the open Air I could not difcern above eight or nine.

O BS.

O B S. XIII.

Appointing an affiftant to move the Prifm to and fro about its Axis, that all the Colours might fuccef-fively fall on that part of the Paper which I faw by reflexion from that part of the Glaffes, where the Circles appeared, fo that all the Colours might be fuccef-fively reflected from the Circles to my Eye whilft I held it immovable, I found the Circles which the red Light made to be manifeftly bigger than thofe which were made by the blue and violet. And it was very plea-fant to fee them gradually fwell or contract according as the Colour of the Light was changed. The inter-val of the Glaffes at any of the Rings when they were made by the utmoft red Light, was to their interval at the fame Ring when made by the utmoft violet, greater than as 3 to 2, and lefs than as 13 to 8, by the moft of my Obfervations it was as 14 to 9. And this proportion feemed very nearly the fame in all obliquities of my Eye; unlefs when two Prifms were made ufe of inftead of the Object-Glaffes. For then at a certain great obliquity of my Eye, the Rings made by the feveral Colours feemed equal, and at a greater obliquity thofe made by the violet would be greater than the fame Rings made by the red. The refraction of the Prifm in this cafe caufing the moft refrangible rays to fall more obliquely on that plate of the Air than the leaft refrangible ones. Thus the Experiment fucceeded in the coloured Light, which was fufficiently ftrong and copious to make the Rings fenfible. And thence it may be gathered, that if the moft refrangible and leaft

refran-

refrangible rays had been copious enough to make the Rings fenfible without the mixture of other rays, the proportion which here was 14 to 9 would have been a little greater, fuppofe 14 $\frac{1}{4}$ or 14 $\frac{1}{3}$ to 9.

O B S. XIV.

Whilft the Prifm was turn'd about its Axis with an uniform motion, to make all the feveral Colours fall fucceffively upon the Object-Glaffes, and thereby to make the Rings contract and dilate : The contraction or dilation of each Ring thus made by the variation of its Colour was fwifteft in the red, and floweft in the violet, and in the intermediate Colours it had inter-mediate degrees of celerity. Comparing the quantity of contraction and dilation made by all the degrees of each Colour, I found that it was greateft in the red ; lefs in the yellow, ftill lefs in the blue, and leaft in the violet. And to make as juft an eftimation as I could of the proportions of their contractions or dilations, I obferved that the whole contraction or dilation of the Diameter of any Ring made by all the degrees of red, was to that of the Diameter of the fame Ring made by all the de-grees of violet, as about four to three, or five to four, and that when the Light was of the middle Colour between yellow and green, the Diameter of the Ring was very nearly an arithmetical mean between the greateft Dia-meter of the fame Ring made by the outmoft red, and the leaft Diameter thereof made by the outmoft violet : Contrary to what happens in the Colours of the oblong Spectrum made by the refraction of a Prifm, where the red is moft contracted, the violet moft expanded, and

D d

in

in the midft of all the Colours is the confine of green and blue. And hence I feem to collect that the thick-neffes of the Air between the Glaffes there, where the Ring is fucceffively made by the limits of the five prin-cipal Colours (red, yellow, green, blue, violet) in order (that is, by the extreme red, by the limit of red and yellow in the middle of the orange, by the limit of yellow and green, by the limit of green and blue, by the limit of blue and violet in the middle of the in-digo, and by the extreme violet) are to one another very nearly as the fix lengths of a Chord which found the notes in a fixth Major, *fol, la, mi, fa, fol, la*. But it agrees fomething better with the Obfervation to fay, that the thickneffes of the Air between the Glaffes there, where the Rings are fucceffively made by the limits of the feven Colours, red, orange, yellow, green, blue, in-digo, violet in order, are to one another as the Cube-roots of the Squares of the eight lengths of a Chord, which found the notes in an eighth, *fol, la, fa, fol, la, mi, fa, fol*; that is, as the Cube-roots of the Squares of the Numbers, $1, \frac{8}{9}, \frac{5}{6}, \frac{3}{4}, \frac{2}{3}, \frac{3}{5}, \frac{9}{16}, \frac{1}{2}$.

O B S. XV.

Thefe Rings were not of various Colours, like thofe made in the open Air, but appeared all over of that prifmatique Colour only with which they were illu-minated. And by projecting the prifmatique Colours immediately upon the Glaffes, I found that the Light which fell on the dark Spaces which were between the coloured Rings, was tranfmitted through the Glaffes without any variation of Colour. For on a
white

white Paper placed behind, it would paint Rings of
the fame Colour with thofe which were reflected, and
of the bignefs of their immediate Spaces. And from
thence the origin of thefe Rings is manifeft ; namely,
That the Air between the Glaffes, according to its various thicknefs, is difpofed in fome places to reflect,
and in others to tranfmit the Light of any one Colour (as you may fee reprefented in the fourth Figure) *Fig.* 4.
and in the fame place to reflect that of one Colour
where it tranfmits that of another.

O B S. XVI.

The Squares of the Diameters of thefe Rings made
by any prifmatique Colour were in arithmetical progreffion as in the fifth Obfervation. And the Diameter
of the fixth Circle, when made by the citrine yellow,
and viewed almoft perpendicularly, was about $\frac{58}{100}$ parts
of an Inch, or a little lefs, agreeable to the fixth Obfervation.

The precedent Obfervations were made with a rarer
thin medium, terminated by a denfer, fuch as was Air
or Water compreffed between two Glaffes. In thofe
that follow are fet down the appearances of a denfer
medium thin'd within a rarer, fuch as are plates of
Mufcovy-glafs, Bubbles of Water, and fome other thin
fubftances terminated on all fides with Air.

　　　　　　　　　　O B S.

O B S. XVII.

If a Bubble be blown with Water firſt made tenacious by diſſolving a little Soap in it, 'tis a common Obſervation, that after a while it will appear tinged with a great variety of Colours. To defend theſe Bubbles from being agitated by the external Air (whereby their Colours are irregularly moved one among another, ſo that no accurate Obſervation can be made of them,) as ſoon as I had blown any of them I covered it with a clear Glaſs, and by that means its Colours emerged in a very regular order, like ſo many concentrick Rings incompaſſing the top of the Bubble. And as the Bubble grew thinner by the continual ſubſiding of the Water, theſe Rings dilated ſlowly and over-ſpread the whole Bubble, deſcending in order to the bottom of it, where they vaniſhed ſucceſſively. In the mean while, after all the Colours were emerged at the top, there grew in the Center of the Rings a ſmall round black Spot, like that in the firſt Obſervation, which continually dilated it ſelf till it became ſometimes more than $\frac{1}{3}$ or $\frac{2}{4}$ of an Inch in breadth before the Bubble broke. At firſt I thought there had been no Light reflected from the Water in that place, but obſerving it more curiouſly, I ſaw within it ſeveral ſmaller round Spots, which appeared much blacker and darker than the reſt, whereby I knew that there was ſome reflexion at the other places which were not ſo dark as thoſe Spots. And by further tryal I fouud that I could ſee the Images of ſome things (as of a Candle or the Sun) very faintly reflected, not only from the great black Spot, but

alſo

Fig. 1.

Fig. 2.

z
z
x
t
r
o
i
g
e
d
c
b

zyxvtsrqponmlkih gfedc   c defghiklmnopqrs tvxyz

Fig. 3.

A   B

C   D

Fig. 4.

A   B
                                                    E
C   D

alſo from the little darker Spots which were within it.

Beſides the aforeſaid coloured Rings there would often appear ſmall Spots of Colours, aſcending and deſcending up and down the ſides of the Bubble, by reaſon of ſome inequalities in the ſubſiding of the Water. And ſometimes ſmall black Spots generated at the ſides would aſcend up to the larger black Spot at the top of the Bubble, and unite with it.

O B S. XVIII.

Becauſe the Colours of theſe Bubbles were more extended and lively than thoſe of the Air thin'd between two Glaſſes, and ſo more eaſy to to diſtinguiſhed, I ſhall here give you a further deſcription of their order, as they were obſerved in viewing them by reflexion of the Skies when of a white Colour, whilſt a black Subſtance was placed behind the Bubble. And they were theſe, red, blue; red, blue; red, blue; red, green; red, yellow, green, blue, purple; red, yellow, green, blue, violet; red, yellow, white, blue, black.

The three firſt Succeſſions of red and blue were very dilute and dirty, eſpecially the firſt, where the red ſeemed in a manner to be white. Among theſe there was ſcarce any other Colour ſenſible beſides red and blue, only the blues (and principally the ſecond blue) inclined a little to green.

The fourth red was alſo dilute and dirty, but not ſo much as the former three; after that ſucceeded little or no yellow, but a copious green, which at firſt inclined a little to yellow, and then became a pretty briſque

and

and good willow green, and afterwards changed to a bluifh Colour; but there fucceeded neither blue nor violet.

The fifth red at firft inclined very much to purple, and afterwards became more bright and brifque, but yet not very pure. This was fucceeded with a very bright and intenfe yellow, which was but little in quantity, and foon changed to green : But that green was copious and fomething more pure, deep and lively, than the former green. After that followed an excellent blue of a bright sky-colour, and then a purple, which was lefs in quantity than the blue, and much inclined to red.

The fixth Red was at firft of a very fair and lively Scarlet, and foon after of a brighter Colour, being very pure and brifque, and the beft of all the reds. Then after a lively orange followed an intenfe bright and copious yellow, which was alfo the beft of all the yellows, and this changed firft to a greenifh yellow, and then to a greenifh blue; but the green between the yellow and the blue, was very little and dilute, feeming rather a greenifh white than a green. The blue which fucceeded became very good, and of a very fair bright sky-colour, but yet fomething inferior to the former blue; and the violet was intenfe and deep with little or no rednefs in it. And lefs in quantity than the blue.

In the laft red appeared a tincture of fcarlet next to violet, which foon changed to a brighter Colour, inclining to an orange; and the yellow which followed was at firft pretty good and lively, but afterwards it grew more dilute, until by degrees it ended in perfect white.

whitenefs. And this whitenefs, if the Water was very tenacious and well-tempered, would flowly fpread and dilate it felf over the greater part of the Bubble; continually growing paler at the top, where at length it would crack in many places, and thofe cracks, as they dilated, would appear of a pretty good, but yet obfcure and dark sky-colour; the white between the blue Spots diminifhing, until it refembled the threds of an irregular Net-work, and foon after vanifhed and left all the upper part of the Bubble of the faid dark blue Colour. And this Colour, after the aforefaid manner, dilated it felf downwards, until fometimes it hath overfpread the whole Bubble. In the mean while at the top, which was of a darker blue than the bottom, and appeared alfo full of many round blue Spots, fomething darker than the reft, there would emerge one or more very black Spots, and within thofe other Spots of an intenfer blacknefs, which I mentioned in the former Obfervation; and thefe continually dilated themfelves until the Bubble broke.

If the Water was not very tenacious the black Spots would break forth in the white, without any fenfible intervention of the blue. And fometimes they would break forth within the precedent yellow, or red, or perhaps within the blue of the fecond order, before the intermediate Colours had time to difplay themfelves.

By this defcription you may perceive how great an affinity thefe Colours have with thofe of Air defcribed in the fourth Obfervation, although fet down in a contrary order, by reafon that they begin to appear when the Bubble is thickeft, and are moft conveniently

niently reckoned from the loweſt and thickeſt part of
the Bubble upwards.

O B S. XIX.

Viewing in ſeveral oblique poſitions of my Eye
the Rings of Colours emerging on the top of the Bubble,
I found that they were ſenſibly dilated by increaſing
the obliquity, but yet not ſo much by far as thoſe
made by thin'd Air in the ſeventh Obſervation. For
there they were dilated ſo much as, when viewed
moſt obliquely, to arrive at a part of the plate more
than twelve times thicker than that where they appeared when viewed perpendicularly; whereas in this
caſe the thickneſs of the Water, at which they arrived
when viewed moſt obliquely, was to that thickneſs
which exhibited them by perpendicular rays, ſomething leſs than as 8 to 5. By the beſt of my Obſervations
it was between 15 and 15½ to 10, an increaſe about
24 times leſs than in the other caſe.

Sometimes the Bubble would become of an uniform
thickneſs all over, except at the top of it near the black
Spot, as I knew, becauſe it would exhibit the ſame
appearance of Colours in all poſitions of the Eye. And
then the Colours which were ſeen at its apparent circumference by the obliqueſt rays, would be different
from thoſe that were ſeen in other places, by rays leſs
oblique to it. And divers Spectators might ſee the
ſame part of it of differing Colours, by viewing it at
very differing obliquities. Now obſerving how much
the Colours at the ſame places of the Bubble, or at divers places of equal thickneſs, were varied by the

ſeveral

feveral obliquities of the rays ; by the affiftance of the
4th, 14th, 16th and 18th Obfervations, as they are
hereafter explained, I colleft the thicknefs of the Water
requifite to exhibit any one and the fame Colour, at feveral obliquities, to be very nearly in the proportion
expreffed in this Table.

Incidence on the Water.		Refraction in- to the Water.		Thickness of the Water.
deg.	min.	deg.	min.	
00	00	00	00	10
15	00	11	11	$10\frac{1}{4}$
30	00	22	1	$10\frac{3}{5}$
45	00	32	2	$11\frac{4}{5}$
60	00	40	30	13
75	00	46	25	$14\frac{1}{2}$
90	00	48	35	$15\frac{1}{5}$

In the two firft Columns are expreffed the obliquities of the rays to the fuperficies of the Water, that
is, their Angles of incidence and refraction. Where
I fuppofe that the Sines which meafure them are in
round numbers as 3 to 4, though probably the diffo-
lution of Soap in the Water, may a little alter its
refractive Vertue. In the third Column the thicknefs
of the Bubble, at which any one Colour is exhibited
in thofe feveral obliquities, is expreft in parts, of which
ten conftitute that thicknefs when the rays are perpendicular.

I have fometimes obferved, that the Colours which
arife on polifhed Steel by heating it, or on Bell-metal,
and fome other metalline fubftances, when melted and

E e poured

poured on the ground, where they may cool in the
open Air, have, like the Colours of Water-bubbles,
been a little changed by viewing them at divers obliquities, and particularly that a deep blue, or violet,
when viewed very obliquely, hath been changed to a
deep red. But the changes of thefe Colours are not fo
great and fenfible as of thofe made by Water. For the
Scoria or vitrified part of the Metal, which moft Metals when heated or melted do continually protrude,
and fend out to their furface, and which by covering
the Metals in form of a thin glaffy skin, caufes thefe
Colours, is much denfer than Water ; and I find that
the change made by the obliquation of the Eye is leaft
in Colours of the denfeft thin fubftances.

O B S. XX.

As in the ninth Obfervation, fo here, the Bubble, by
tranfmitted Light, appeared of a contrary Colour to
that which it exhibited by reflexion. Thus when the
Bubble being looked on by the Light of the Clouds reflected from it, feemed red at its apparent circumference, if the Clouds at the fame time, or immediately
after, were viewed through it, the Colour at its circumference would be blue. And, on the contrary,
when by reflected Light it appeared blue, it would appear red by tranfmitted Light.

O B S. XXI.

By wetting very thin plates of Mufcovy-glafs, whofe
thinnefs made the like Colours appear, the Colours
became

became more faint and languid; especially by wetting the plates on that side oppofite to the Eye: But I could not perceive any variation of their fpecies. So then the thicknefs of a plate requifite to produce any Colour, depends only on the denfity of the plate, and not on that of the ambient medium: And hence, by the 10th and 16th Obfervations, may be known the thicknefs which Bubbles of Water, or Plates of Mufcovy-glafs, or other fubftances, have at any Colour produced by them.

O B S. XXII.

A thin tranfparent Body, which is denfer than its ambient medium, exhibits more brifque and vivid Colours than that which is fo much rarer; as I have particularly obferved in the Air and Glafs. For blowing Glafs very thin at a Lamp-furnace, thofe plates incompaffed with Air did exhibit Colours much more vivid than thofe of Air made thin between two Glaffes.

O B S. XXIII.

Comparing the quantity of Light reflected from the feveral Rings, I found that it was moft copious from the firft or inmoft, and in the exterior Rings became gradually lefs and lefs. Alfo the whitenefs of the firft Ring was ftronger than that reflected from thofe parts of the thinner medium which were without the Rings; as I could manifeftly perceive by viewing at a diftance the Rings made by the two Object-

Glaffes,

Glaffes; or by comparing two Bubbles of Water blown
at diftant times, in the firft of which the whitenefs
appeared, which fucceeded all the Colours, and in,
the other, the whitenefs which preceded them all.

O B S. XXIV.

When the two Object-Glaffes were lay'd upon one
another, fo as to make the Rings of the Colours appear, though with my naked Eye I could not difcern
above 8 or 9 of thofe Rings, yet by viewing them
through a Prifm I have feen a far greater multitude,
infomuch that I could number more than forty, befides
many others, that were fo very fmall and clofe together, that I could not keep my Eye fteddy on them
feverally fo as to number them, but by their extent I have
fometimes eftimated them to be more than a hundred.
And I believe the Experiment may be improved to the
difcovery of far greater numbers. For they feem to
be really unlimited, though vifible only fo far as they
can be feparated by the refraction, as I fhall hereafter
explain.

But it was but one fide of thefe Rings, namely, that
towards which the refraction was made, which by that
refraction was rendered diftinct, and the other fide became more confufed than when viewed by the naked
Eye, infomuch that there I could not difcern above
one or two, and fometimes none of thofe Rings, of
which I could difcern eight or nine with my naked
Eye. And their Segments or Arcs, which on the
other fide appeared fo numerous, for the moft part
exceeded

exceeded not the third part of a Circle. If the Refraction was very great, or the Prifm very diftant from the Object-Glaffes, the middle part of thofe Arcs became alfo confufed, fo as to difappear and conftitute an even whitenefs, whilft on either fide their ends, as alfo the whole Arcs furtheft from the center, became diftincter than before, appearing in the form as you fee them defigned in the fifth Figure. *Fig.* 5.

The Arcs, where they feemed diftincteft, were only white and black fucceffively, without any other Colours intermixed. But in other places there appeared Colours, whofe order was inverted by the refraction in fuch manner, that if I firft held the Prifm very near the Object-Glaffes, and then gradually removed it further off towards my Eye, the Colours of the 2d, 3d, 4th, and following Rings fhrunk towards the white that emerged between them, until they wholly vanifhed into it at the middle of the Arcs, and afterwards emerged again in a contrary order. But at the ends of the Arcs they retained their order unchanged.

I have fometimes fo lay'd one Object-Glafs upon the other, that to the naked Eye they have all over feemed uniformly white, without the leaft appearance of any of the coloured Rings ; and yet by viewing them through a Prifm, great multitudes of thofe Rings have difcovered themfelves. And in like manner plates of Mufcovy-glafs, and Bubbles of Glafs blown at a Lamp-furnace, which were not fo thin as to exhibit any Colours to the naked Eye, have through the Prifm exhibited a great variety of them ranged irregularly up and down in the form of waves. And fo

Bubbles

Bubbles of Water, before they began to exhibit their Colours to the naked Eye of a By-ſtander, have appeared through a Priſm, girded about with many parallel and horizontal Rings; to produce which effect, it was neceſſary to hold the Priſm parallel, or very nearly parallel to the Horizon, and to diſpoſe it ſo that the rays might be refracted upwards.

THE

THE
SECOND BOOK
OF
OPTICKS.

PART II.

Remarks upon the foregoing Observations.

HAving given my Observations of these Colours, before I make use of them to unfold the Causes of the Colours of natural Bodies, it is convenient that by the simplest of them, such as are the 2d, 3d, 4th, 9th, 12th, 18th, 20th, and 24th, I first explain the more expounded. And first to shew how the Colours in the fourth and eighteenth Observations are produced, let there be taken in any right line from the point Y, the lengths YA, YB, YC, YD, YE, YF, YG, *Fig.5.* YH, in proportion to one another, as the Cube-roots of the Squares of the numbers, $\frac{1}{2}$, $\frac{9}{16}$, $\frac{3}{5}$, $\frac{2}{3}$, $\frac{3}{4}$, $\frac{5}{6}$, $\frac{8}{9}$, 1, whereby the lengths of a musical Chord to found all the Notes in an Eighth are represented; that is, in the proportion of the numbers 6300, 6814, 7114, 7631, 8255, 8855, 9243, 10000. And at the points A, B, C, D, E, F,

E, F, G, H, let perpendiculars A α, B β, &c. be erected, by whose intervals the extent of the several Colours set underneath against them, is to be represented. Then divide the line A α in such proportion as the numbers 1, 2, 3, 5, 6, 7, 9, 10, 11, &c. set at the points of division denote. And through those divisions from Y draw lines 1 I, 2 K, 3 L, 5 M, 6 N, 7 O, &c.

Now if A 2 be supposed to represent the thickness of any thin transparent Body, at which the outmost violet is most copiously reflected in the first Ring, or Series of Colours, then by the 13th Observation H K, will represent its thickness, at which the utmost red is most copiously reflected in the same Series. Also by the 5th and 16th Observations, A 6 and H N will denote the thicknesses at which those extreme Colours are most copiously reflected in the second Series, and A 10 and H Q the thicknesses, at which they are most copiously reflected in the third Series, and so on. And the thickness at which any of the intermediate Colours are reflected most copiously, will, according to the 14th Observation, be defined by the distance of the line A H from the intermediate parts of the lines 2 K, 6 N, 10 Q, &c. against which the names of those Colours are written below.

But further, to define the latitude of these Colours in each Ring or Series, let A 1 design the least thickness, and A 3 the greatest thickness, at which the extreme violet in the first Series is reflected, and let H I, and H L, design the like limits for the extreme red, and let the intermediate Colours be limited by the intermediate parts of the lines 1 I, and 3 L, against which the names of those Colours are written, and so on: But
yet

yet with this caution, that the reflections be suppofed strongeft at the intermediate Spaces, 2 K, 6 N, 10 Q, &c. and from thence to decreafe gradually towards thefe limits, 1 I, 3 L, 5 M, 7 O, &c. on either fide; where you muft not conceive them to be precifely limited, but to decay indefinitely. And whereas I have affigned the fame latitude to every Series, I did it, becaufe although the Colours in the firft Series feem to be a little broader than the reft, by reafon of a ftronger reflexion there, yet that inequality is fo infenfible as fcarcely to be determined by Obfervation.

Now according to this defcription, conceiving that the rays originally of feveral Colours are by turns reflected at the Spaces 1 I L 3, 5 M O 7, 9 P R 11, &c. and tranfmitted at the Spaces A H I 1, 3 L M 5, 7 O P 9, &c. it is eafy to know what Colour muft in the open Air be exhibited at any thicknefs of a tranfparent thin body. For if a Ruler be applied parallel to A H, at that diftance from it by which the thicknefs of the body is reprefented, the alternate Spaces 1 I L 3, 5 M O 7, &c. which it croffeth will denote the reflected original Colours, of which the Colour exhibited in the open Air is compounded. Thus if the conftitution of the green in the third Series of Colours be defired, apply the Ruler as you fee at $\pi\, \varrho\, \sigma\, \varphi$, and by its paffing through fome of the blue at π and yellow at σ, as well as through the green at ϱ, you may conclude that the green exhibited at that thicknefs of the body is principally conftituted of original green, but not without a mixture of fome blue and yellow.

F f

By

By this means you may know how the Colours from
the center of the Rings outward ought to fucceed in
order as they were defcribed in the 4th and 18th Obfervations. For if you move the Ruler gradually from
A H through all diftances, having paft over the firft
fpace which denotes little or no reflexion to be made
by thinneft fubftances, it will firft arrive at 1 the violet,
and then very quickly at the blue and green, which
together with that violet compound blue, and then at
the yellow and red , by whofe further addition that
blue is converted into whitenefs, which whitenefs continues during the tranfit of the edge of the Ruler from
I to 3, and after that by the fucceffive deficience of
its component Colours, turns firft to compound yellow,
and then to red, and laft of all the red ceafeth at L.
Then begin the Colours of the fecond Series, which
fucceed in order during the tranfit of the edge of the
Ruier from 5 to O, and are more lively than before,
becaufe more expanded and fevered. And for the
fame reafon, inftead of the former white there intercedes between the blue and yellow a mixture of orange,
yellow, green, blue and indico, all which together ought
to exhibit a dilute and imperfect green. So the Colours of the third Series all fucceed in order ; firft, the
violet, which a little interferes with the red of the fecond order, and is thereby inclined to a reddifh purple;
then the blue and green, which are lefs mixed with
other Coiours, and confequently more lively than before, efpecially the green : Then follows the yellow,
fome of which towards the green is diftinct and good, but
that part of it towards the fucceeding red, as alfo that
red is mixed with the violet and blue of the fourth Series,

ries, whereby various degrees of red very much inclining to purple are compounded. This violet and blue, which should succeed this red, being mixed with, and hidden in it, there succeeds a green. And this at first is much inclined to blue, but soon becomes a good green, the only unmixed and lively Colour in this fourth Series. For as it verges towards the yellow, it begins to interfere with the Colours of the fifth Series, by whose mixture the succeeding yellow and red are very much diluted and made dirty, especially the yellow, which being the weaker Colour is scarce able to shew it self. After this the several Series interfere more and more, and their Colours become more and more intermixed, till after three or four more revolutions (in which the red and blue predominate by turns) all sorts of Colours are in all places pretty equally bended, and compound an even whiteness.

And since by the 15th Observation the rays indued with one Colour are transmitted, where those of another Colour are reflected, the reason of the Colours made by the transmitted Light in the 9th and 20th Observations is from hence evident.

If not only the order and species of these Colours, but also the precise thickness of the plate, or thin body at which they are exhibited, be desired in parts of an Inch, that may be also obtained by assistance of the 6th or 16th Observations. For according to those Observations the thickness of the thinned Air, which between two Glasses exhibited the most luminous parts of the first six Rings were $\frac{1}{178000}$, $\frac{3}{178000}$, $\frac{5}{178000}$, $\frac{7}{178000}$, $\frac{9}{178000}$, $\frac{11}{178000}$ parts of an Inch. Suppose the Light reflected most copiously at these thicknesses be the bright citrine yellow, or con-

fine

fine of yellow and orange, and thefe thickneffes will be $G\mu$, $G\nu$, $G\xi$, Go, $G\eta$. And this being known, it is eafy to determine what thicknefs of Air is reprefented by $G\varphi$, or by any other diftance of the ruler from A H.

But further, fince by the 10th Obfervation the thicknefs of Air was to the thicknefs of Water, which between the fame Glaffes exhibited the fame Colour, as 4 to 3, and by the 21th Obfervation the Colours of thin bodies are not varied by varying the ambient medium ; the thicknefs of a Bubble of Water, exhibiting any Colour, will be $\frac{3}{4}$ of the thicknefs of Air producing the fame Colour. And fo according to the fame 10th and 21th Obfervations the thicknefs of a plate of Glafs, whofe refraction of the mean refrangible ray, is meafured by the proportion of the Sines 31 to 20, may be $\frac{20}{31}$ of the thicknefs of Air producing the fame Colours; and the like of other mediums. I do not affirm, that this proportion of 20 to 31, holds in all the rays ; for the Sines of other forts of rays have other proportions. But the differences of thofe proportions are fo little that I do not here confider them. On thefe Grounds I have compofed the following Table, wherein the thicknefs of Air, Water, and Glafs, at which each Colour is moft intenfe and fpecifick, is expreffed in parts of an Inch divided into Ten hundred thoufand equal parts.

The

The thickness of coloured Plates and Particles of

		Air.	Water.	Glass.
Their Colours of the first Order,	Very Black	$\frac{1}{2}$	$\frac{1}{8}$	$\frac{10}{31}$
	Black	1	$\frac{1}{4}$	$\frac{40}{31}$
	Beginning of Black	2	$1\frac{1}{2}$	$1\frac{2}{7}$
	Blue	$2\frac{2}{5}$	$1\frac{4}{5}$	$1\frac{11}{20}$
	White	$5\frac{1}{4}$	$3\frac{3}{8}$	$3\frac{3}{5}$
	Yellow	$7\frac{1}{9}$	$5\frac{1}{3}$	$4\frac{3}{5}$
	Orange	8	6	$5\frac{5}{6}$
	Red	9	$6\frac{1}{4}$	$5\frac{4}{5}$
Of the second Order,	Violet	$11\frac{1}{6}$	$8\frac{3}{8}$	$7\frac{1}{3}$
	Indico	$12\frac{1}{6}$	$9\frac{1}{8}$	$8\frac{1}{11}$
	Blue	14	$10\frac{1}{2}$	9
	Green	$15\frac{1}{8}$	$11\frac{1}{3}$	$9\frac{1}{7}$
	Yellow	$16\frac{2}{7}$	$12\frac{1}{5}$	$10\frac{3}{5}$
	Orange	$17\frac{2}{9}$	13	$11\frac{1}{9}$
	Bright Red	$18\frac{1}{3}$	$13\frac{1}{4}$	$11\frac{1}{6}$
	Scarlet	$19\frac{1}{3}$	$14\frac{1}{4}$	$12\frac{2}{3}$
Of the third Order,	Purple	21	$15\frac{3}{4}$	$13\frac{11}{20}$
	Indico	$22\frac{1}{10}$	$16\frac{4}{7}$	$14\frac{1}{4}$
	Blue	$23\frac{3}{5}$	$17\frac{11}{20}$	$15\frac{1}{10}$
	Green	$25\frac{1}{5}$	$18\frac{9}{10}$	$16\frac{1}{4}$
	Yellow	$27\frac{1}{7}$	$20\frac{1}{3}$	$17\frac{1}{2}$
	Red	29	$21\frac{1}{4}$	$18\frac{1}{7}$
	Bluish Red	32	24	$20\frac{2}{3}$
Of the fourth Order,	Bluish Green	34	$25\frac{1}{2}$	22
	Green	$35\frac{2}{7}$	$26\frac{1}{2}$	$22\frac{1}{4}$
	Yellowish Green	36	27	$23\frac{2}{9}$
	Red	$40\frac{1}{3}$	$30\frac{1}{4}$	26
Of the fifth Order,	Greenish Blue	46	$34\frac{1}{2}$	$29\frac{2}{3}$
	Red	$52\frac{1}{2}$	$39\frac{3}{8}$	34
Of the sixth Order,	Greenish Blue	$58\frac{1}{4}$	44	38
	Red	65	$48\frac{1}{4}$	42
Of the seventh Order,	Greenish Blue	71	$53\frac{3}{4}$	$45\frac{2}{5}$
	Ruddy White	77	$57\frac{3}{4}$	$49\frac{1}{3}$

Now

Now if this Table be compared with the 6th Scheme, you will there fee the conftitution of each Colour, as to its Ingredients, or the original Colours of which it is compounded, and thence be enabled to judge of its intenfenefs or imperfection; which may fuffice in ex-plication of the 4th and 18th Obfervations, unlefs it be further defired to delineate the manner how the Co-lours appear, when the two Object-Glaffes are lay'd upon one another. To do which, let there be de-fcribed a large Arc of a Circle, and a ftreight Line which may touch that Arc, and parallel to that Tan-gent feveral occult Lines, at fuch diftances from it, as the numbers fet againft the feveral Colours in the Table denote. For the Arc, and its Tangent, will reprefent the fuperficies of the Glaffes terminating the interjacent Air; and the places where the occult Lines cut the Arc will fhow at what diftances from the Center, or Point of contact, each Colour is reflected.

There are alfo other ufes of this Table : For by its affiftance the thicknefs of the Bubble in the 19th Ob-fervation was determined by the Colours which it ex-hibited. And fo the bignefs of the parts of natural Bodies may be conjectured by their Colours, as fhall be hereafter fhewn. Alfo, if two or more very thin plates be lay'd one upon another, fo as to compofe one plate equalling them all in thicknefs, the refulting Colour may be hereby determined. For inftance, Mr. *Hook* in his *Mifcrographia* obferves, that a faint yellow plate of Mufcovy-glafs lay'd upon a blue one, conftituted a very deep purple. The yellow of the firft Order is a faint one, and the thicknefs of the plate exhibiting it, ac-cording to the Table is 4$\frac{3}{5}$, to which add 9, the thick-

nefs

nefs exhibiting blue of the fecond Order, and the fum will be $13\frac{1}{9}$, which is the thicknefs exhibiting the purple of the third Order.

To explain, in the next place, the Circumftances of the 2d and 3d Obfervations; that is, how the Rings of the Colours may (by turning the Prifms about their common Axis the contrary way to that expreffed in thofe Obfervations) be converted into white and black Rings, and afterwards into Rings of Colours again, the Colours of each Ring lying now in an inverted order; it muft be remembred, that thofe Rings of Colours are dilated by the obliquation of the rays to the Air which intercedes the Glaffes, and that according to the Table in the 7th Obfervation, their dilatation or increafe of their Diameter is moft manifeft and fpeedy when they are obliqueft. Now the rays of yellow being more refracted by the firft fuperficies of the faid Air than thofe of red, are thereby made more oblique to the fecond fuperficies, at which they are reflected to produce the coloured Rings, and confequently the yellow Circle in each Ring will be more dilated than the red; and the excefs of its dilatation will be fo much the greater, by how much the greater is the obliquity of the rays, until at laft it become of equal extent with the red of the fame Ring. And for the fame reafon the green, blue and violet, will be alfo fo much dilated by the ftill greater obliquity of their rays, as to become all very nearly of equal extent with the red, that is, equally diftant from the center of the Rings. And then all the Colours of the fame Ring muft be coincident, and by their mixture exhibit a white Ring. And thefe white Rings muft have black and dark Rings between them, becaufe they do not

fpread

spread and interfere with one another as before. And
for that reason also they muft become diftinёler and vifible to far greater Numbers. But yet the violet being
obliqueft will be fomething more dilated in proportion
to its extent then the other Colours, and fo very apt to
appear at the exterior verges of the white.

Afterwards, by a greater obliquity of the rays, the
violet and blue become more fenfibly dilated than the
red and yellow, and fo being further removed from the
center of the Rings, the Colours muft emerge out of the
white in an order contrary to that which they had before, the violet and blue at the exterior limbs of each
Ring, and the red and yellow at the interior. And the violet, by reafon of the greateft obliquity of its rays, being
in proportion moft of all expanded, will fooneft appear
at the exterior limb of each white Ring, and become
more confpicuous than the reft. And the feveral Series
of Colours belonging to the feveral Rings, will, by
their unfolding and fpreading, begin again to interfere,
and thereby render the Rings lefs diftinёt, and not vifible to fo great numbers.

If inftead of the Prifms the Objeёt-glaffes be made
ufe of, the Rings which they exhibit become not white
and diftinёt by the obliquity of the Eye, by reafon that
the rays in their paffage through that Air which intercedes the Glaffes are very nearly parallel to thofe Lines
in which they were firft incident on the Glaffes, and confequently the rays indued with feveral Colours are not
inclined one more than another to that Air, as it happens in the Prifms.

There is yet another circumftance of thefe Experiments
to be confidered, and that is why the black and white
Rings

Rings which when viewed at a diftance appear diftinct, fhould not only become confufed by viewing them near at hand , but alfo yield a violet Colour at both the edges of every white Ring. And the reafon is, that the rays which enter the Eye at feveral parts of the Pupil, have feveral obliquities to the Glaffes, and thofe which are moft oblique, if confidered apart, would reprefent the Rings bigger than thofe which are the leaft oblique. Whence the breadth of the perimeter of every white Ring is expanded outwards by the obliqueft rays, and inwards by the leaft oblique. And this expanfion is fo much the greater by how much the greater is the difference of the obliquity ; that is, by how much the Pupil is wider, or the Eye nearer to the Glaffes. And the breadth of the violet muft be moft expanded, be- caufe the rays apt to excite a fenfation of that Colour are moft oblique to a fecond, or further fuperficies of the thin'd Air at which they are reflected, and have alfo the greateft variation of obliquity, which makes that Colour fooneft emerge out of the edges of the white. And as the breadth of every Ring is thus aug- mented, the dark intervals muft be diminifhed, until the neighbouring Rings become continuous, and are blended, the exterior firft, and then thofe nearer the Center, fo that they can no longer be diftinguifh'd apart, but feem to conftitute an even and uniform whitenefs.

Among all the Obfervations there is none accompa- nied with fo odd circumftances as the 24th. Of thofe the principal are, that in thin plates, which to the naked Eye feem of an even and uniform tranfparent

G g   white-

whitenefs, without any terminations of fhadows, the refraction of a Prifm fhould make Rings of Colours appear, whereas it ufually makes Objects appear coloured only there where they are terminated with fhadows, or have parts unequally luminous; and that it fhould make thofe Rings exceedingly diftinct and white, although it ufually renders Objects confufed and coloured. The caufe of thefe things you will underftand by confidering, that all the Rings of Colours are really in the plate, when viewed with the naked Eye, although by reafon of the great breadth of their circumferences they fo much interfere and are blended together, that they feem to conftitute an even whitenefs. But when the rays pafs through the Prifm to the Eye, the orbits of the feveral Colours in every Ring are refracted, fome more than others, according to their degrees of refrangibility: By which means the Colours on one fide of the Ring (that is on one fide of its Center) become more unfolded and dilated, and thofe on the other fide more complicated and contracted. And where by a due refraction they are fo much contracted, that the fevral Rings become narrower than to interfere with one another, they muft appear diftinct, and alfo white, if the conftituent Colours be fo much contracted as to be wholly coincident. But, on the other fide, where the orbit of every Ring is made broader by the further unfolding of its Colours, it muft interfere more with other Rings than before, and fo become lefs diftinct.

To explain this a little further, fuppofe the concentrick Circles A V, and B X, reprefent the red and violet of any order, which, together with the intermediate Colours,

Fig. 7.

Colours, conſtitute any one of theſe Rings. Now theſe being viewed through a Priſm, the violet Circle B X, will by a greater refraction be further tranſlated from its place than the red A V, and ſo approach nearer to it on that ſide, towards which the refractions are made. For inſtance, if the red be tranſlated to *av*, the violet may be tranſlated to *b x*, ſo as to approach nearer to it at *x* than before, and if the red be further tranſlated to *a* v, the violet may be ſo much further tranſlated to b x as to convene with it at x, and if the red be yet further tranſlated to *α* ϒ, the violet may be ſtill ſo much further tranſlated to *β* ξ as to paſs beyond it at ξ, and convene with it at *e* and *f*. And this being underſtood not only of the red and violet, but of all the other intermediate Colours, and alſo of every revolution of thoſe Colours, you will eaſily perceive how thoſe of the ſame revolution or order, by their nearneſs at *x v* and ϒ ξ, and their coincidence at x v, *e* and *f*, ought to conſtitute pretty diſtinct Arcs of Circles, eſpecially at x v, or at *e* and *f*, and that they will appear ſeverally at *x v*, and at x v exhibit whiteneſs by their coincidence, and again appear ſeveral at ϒ ξ, but yet in a contrary order to that which they had before, and ſtill retain beyond *e* and *f*. But, on the other ſide, at *ab*, a b, or *α β*, theſe Colours muſt become much more confuſed by being dilated and ſpread ſo, as to interfere with thoſe of other Orders. And the ſame confuſion will happen at ϒ ξ between *e* and *f*, if the refraction be very great, or the Priſm very diſtant from the Object-Glaſſes: In which caſe no parts of the Rings will be ſeen, ſave only two little Arcs at *e* and *f*, whoſe diſtance from one

another,

another will be augmented by removing the Prifm ftill further from the Object-Glaffes: And thefe little Arcs muft be diftincteft and whiteft at their middle, and at their ends, where they begin to grow confufed they muft be coloured. And the Colours at one end of every Arc muft be in a contrary order to thofe at the other end, by reafon that they crofs in the intermediate white; namely their ends, which verge towards $\Upsilon\xi$, will be red and yellow on that fide next the Center, and blue and violet on the other fide. But their other ends which verge from $\Upsilon\xi$ will on the contrary be blue and violet on that fide towards the Center, and on the other fide red and yellow.

Now as all thefe things follow from the Properties of Light by a mathematical way of reafoning, fo the truth of them may be manifefted by Experiments. For in a dark room, by viewing thefe Rings through a Prifm, by reflexion of the feveral prifmatique Colours, which an affiftant caufes to move to and fro upon a Wall or Paper from whence they are reflected, whilft the Spectator's Eye, the Prifm and the Object-Glaffes (as in the 13th Obfervation) are placed fteddy: the pofition of the Circles made fucceffively by the feveral Colours, will be found fuch, in refpect of one another, as I have defcribed in the Figures $abxv$, or $abxv$, or $\alpha\beta\xi\Upsilon$. And by the fame method the truth of the Explications of other Obfervations may be examined.

By what hath been faid the like Phænomina of Water, and thin plates of Glafs may be underftood. But in fmall fragments of thofe plates, there is this

further

further obfervable, that where they lye flat upon a
Table and are turned about their Centers whilft they are
viewed through a Prifm, they will in fome poftures
exhibit waves of various Colours, and fome of them exhibit thefe waves in one or two pofitions only, but the
moft of them do in all pofitions exhibit them, and make
them for the moft part appear almoft all over the plates.
The reafon is, that the fuperficies of fuch plates are not
even, but have many cavities and fwellings, which how
fhallow foever do a little vary the thicknefs of the
plate. For at the feveral fides of thofe cavities, for
the reafons newly defcribed, there ought to be produced waves in feveral poftures of the Prifm. Now though
it be but fome very fmall, and narrower parts of the
Glafs, by which thefe waves for the moft part are caufed, yet they may feem to extend themfelves over the
whole Glafs, becaufe from the narroweft of thofe parts
there are Colours of feveral Orders that is of feveral
Rings, confufedly reflected, which by refraction of the
Prifm are unfolded, feparated, and according to their
degrees of refraction, difperfed to feveral places, fo as to
conftitute fo many feveral waves, as there were divers
orders of Colours promifcuoufly reflected from that
part of the Glafs.

Thefe are the principal Phænomena of thin Plates
or Bubbles, whofe explications depend on the properties of Light, which I have heretofore delivered.
And thefe you fee do neceffarily follow from them, and
agree with them, even to their very leaft circumftances;
and not only fo, but do very much tend to their proof.
Thus, by the 24th Obfervation, it appears, that the

rays

rays of feveral Colours made as well by thin Plates or Bubbles, as by refractions of a Prifm, have feveral degrees of refrangibility, whereby thofe of each order, which at their reflexion from the Plate or Bubble are intermixed with thofe of other orders, are feparated from them by refraction, and affociated together fo as to become vifible by themfelves like Arcs of Circles. For if the rays were all alike refrangible, 'tis impoffible that the whitenefs, which to the naked fence appears uniform, fhould by refraction have its parts tranfpofed and ranged into thofe black and white Arcs.

It appears alfo that the unequal refractions of difform rays proceed not from any contingent irregularities ; fuch as are veins, an uneven polifh, or fortuitous pofition of the pores of Glafs ; unequal and cafual motions in the Air or Æther ; the fpreading, breaking, or dividing the fame ray into many diverging parts, or the like. For, admitting any fuch irregularities, it would be impoffible for refractions to render thofe Rings fo very diftinct, and well defined, as they do in the 24th Obfervation. It is neceffary therefore that every ray have its proper and conftant degree of refrangibility connate with it, according to which its refraction is ever juftly and regularly performed, and that feveral rays have feveral of thofe degrees.

And what is faid of their refrangibility may be alfo underftood of their reflexibility, that is of their difpofitions to be reflected fome at a greater, and others at a lefs thicknefs, of thin Plates or Bubbles, namely, that thofe difpofitions are alfo connate with the rays, and immutable ; as may appear by the 13th, 14th, and
15th

15th Obfervations compared with the fourth and eighth.

By the precedent Obfervations it appears alfo, that whitenefs is a diffimilar mixture of all Colours, and that Light is a mixture of rays indued with all thofe Colours. For confidering the multitude of the Rings of Colours, in the 3d, 12th and 24th Obfervations, it is manifeft that although in the 4th and 18th Obfervations there appear no more than eight or nine of thofe Rings, yet there are really a far greater number, which fo much interfere and mingle with one another, as after thofe eight or nine revolutions to dilute one another wholly, and conftitute an even and fenfibly uniform whitenefs. And confequently that whitenefs muft be allowed a mixture of all Colours, and the Light which conveys it to the Eye muft be a mixture of rays indued with all thofe Colours.

But further, by the 24th Obfervation, it appears, that there is a conftant relation between Colours and Refrangibility, the moft refrangible rays being violet, the leaft refrangible red, and thofe of intermediate Colours having proportionably intermediate degrees of refrangibility. And by the 13th, 14th and 15th Obfervations, compared with the 4th or 18th, there appears to be the fame conftant relation between Colour and Reflexibility, the violet being in like circumftances reflected at leaft thickneffes of any thin Plate or Bubble, the red at greateft thickneffes, and the intermediate Colours at intermediate thickneffes. Whence it follows, that the colorifique difpofitions of rays are alfo connate with them and immutable, and by confequence

that

that all the productions and appearances of Colours in the World are derived not from any phyfical change caufed in Light by refraction or reflexion, but only from the various mixtures or feparations of rays, by virtue of their different Refrangibility or Reflexibility. And in this refpect the Science of Colours becomes a Speculation as truly mathematical as any other part of Optiques. I mean fo far as they depend on the nature of Light, and are not produced or altered by the power of imagination, or by ftriking or preffing the Eyes.

THE

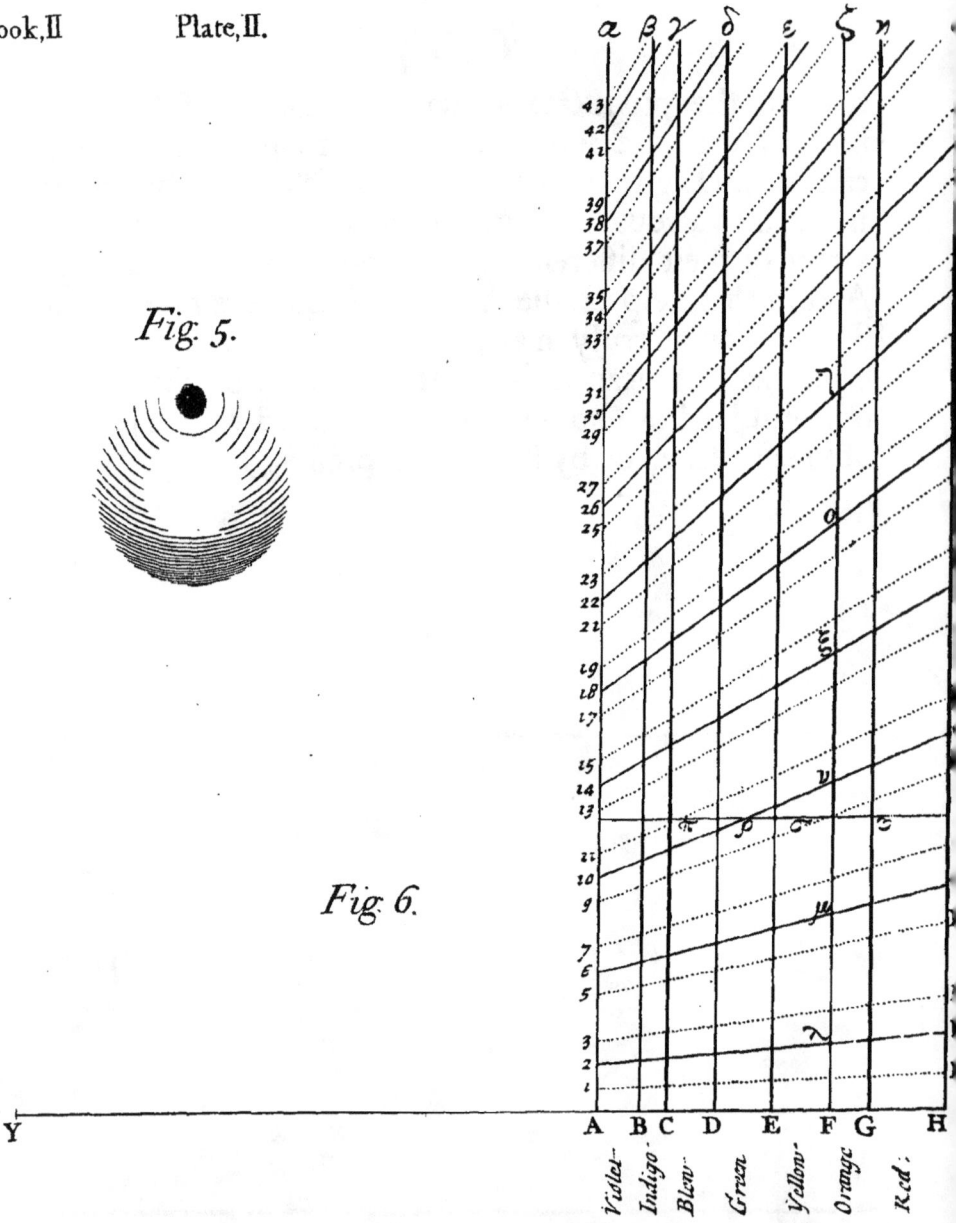

a β γ δ ϵ ζ η

43
42
41

39
38
37

35
34
33

31
30
29

27
26
25

23
22
21

19
18
17

15
14
13

11
10
9

7
6
5

3
2
1

Y   A B C D   E   F G   H

Violet *Indigo* *Bleu* *Green* *Yellow* *Orange* *Red*

Fig. 5.

Fig. 6.

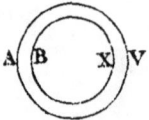

Fig. 7.

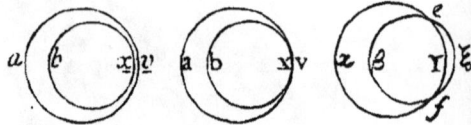

THE

SECOND BOOK

OF

OPTICKS.

PART III.

Of the permanent Colours of natural Bodies, and the Analogy between them and the Colours of thin transparent Plates.

I Am now come to another part of this Design, which is to consider how the Phænomena of thin transparent Plates stand related to those of all other natural Bodies. Of these Bodies I have already told you that they appear of divers Colours, accordingly as they are disposed to reflect most copiously the rays originally indued with those Colours. But their Constitutions, whereby they reflect some rays more copiously than others, remains to be discovered, and these I shall endeavour to manifest in the following Propositions.

H h   PROP.

PROP. I.

Those superficies of transparent Bodies reflect the greatest quantity of Light, which have the greatest refracting power; that is, which intercede mediums that differ most in their refractive densities. And in the confines of equally refracting mediums there is no reflexion.

The Analogy between reflexion and refraction will appear by considering, that when Light passeth obliquely out of one medium into another which refracts from the perpendicular, the greater is difference of their refractive density, the less obliquity is requisite to cause a total reflexion. For as the Sines are which measure the refraction, so is the Sine of incidence at which the total reflexion begins, to the radius of the Circle, and consequently that incidence is least where there is the greatest difference of the Sines. Thus in the passing of Light out of Water into Air, where the refraction is measured by the Ratio of the Sines 3 to 4, the total reflexion begins when the Angle of incidence is about 48 degrees 35 minutes. In passing out of Glass into Air, where the refraction is measured by the Ratio of the Sines 20 to 31, the total reflexion begins when the Angle of incidence is 40 deg. 10 min. and so in passing out of cryſtal, or more ſtrongly refracting mediums into Air, there is ſtill a leſs obliquity requiſite to cauſe a total reflexion. Superficies therefore which refract moſt do ſooneſt reflect all the Light which is incident on them, and ſo muſt be allowed moſt ſtrongly reflexive.

But

But the truth of this Propofition will further appear by obferving, that in the fuperficies interceding two tranfparent mediums, fuch as are (Air, Water, Oyl, Common-Glafs, Cryftal, Metalline-Glaffes, Ifland-Glaffes, white tranfparent Arfnick, Diamonds, &c.) the reflexion is ftronger or weaker accordingly, as the fuperficies hath a greater or lefs refracting power. For in the confine of Air and Sal-gemm 'tis ftronger than in the confine of Air and Water, and ftill ftronger in the confine of Air and Common-Glafs or Cryftal, and ftronger in the confine of Air and a Diamond. If any of thefe, and fuch like tranfparent Solids, be immerged in Water, its reflexion becomes much weaker than before, and ftill weaker if they be immerged in the more ftrongly refracting Liquors of well-rectified oyl of Vitriol or fpirit of Turpentine. If Water be diftinguifhed into two parts, by any imaginary furface, the reflexion in the confine of thofe two parts is none at all. In the confine of Water and Ice 'tis very little, in that of Water and Oyl 'tis fomething greater, in that of Water and Sal-gemm ftill greater, and in that of Water and Glafs, or Cryftal, or other denfer fubftances ftill greater, accordingly as thofe mediums differ more or lefs in their refracting powers. Hence in the confine of Common-Glafs and Cryftal, there ought to be a weak reflexion, and a ftronger reflexion in the confine of Common and Metalline-Glafs, though I have not yet tried this. But, in the confine of two Glaffes of equal denfity, there is not any fenfible reflexion, as was fhewn in the firft Obfervation. And the fame may be underftood of the fuperficies interceding two Cryftals, or two Liquors, or any other Subftances in which no refraction is caufed. So then the

reafon

reafon why uniform pellucid mediums, (fuch as Water, Glafs, or Cryftal) have no fenfible reflexion but in their external fuperficies, where they are adjacent to other mediums of a different denfity, is becaufe all their contiguous parts have one and the fame degree of denfity.

PROP. II.

The leaft parts of almoft all natural Bodies are in fome meafure tranfparent : And the opacity of thofe Bodies arifeth from the multitude of reflexions caufed in their internal Parts.

That this is fo has been obferved by others, and will eafily be granted by them that have been converfant with Mifcrofcopes. And it may be alfo tryed by applying any fubftance to a Hole through which fome Light is immitted into a dark room. For how opake foever that fubftance may feem in the open Air, it will by that means appear very manifeftly tranfparent, if it be of a fufficient thinnefs. Only white metalline Bodies muft be excepted, which by reafon of their exceffive denfity feem to reflect almoft all the Light incident on their firft fuperficies, unlefs by folution in menftruums they be reduced into very fmall particles, and then they become tranfparent.

PROP. III.

Between the parts of opake and coloured Bodies are many fpaces, either empty or replenifhed, with mediums of other denfities ; as Water between the tinging corpufcles wherewith any Liquor is impregnated, Air between the

aqueous

aqueous globules that constitute Clouds or Mists ; and for the most part spaces void of both Air and Water, but yet perhaps not wholly void of all substance, between the parts of hard Bodies.

The truth of this is evinced by the two precedent Propofitions : For by the fecond Propofition there are many reflexions made by the internal parts of Bodies, which, by the firft Propofition, would not happen if the parts of thofe Bodies were continued without any fuch interftices between them, becaufe reflexions are caufed only in fuperficies, which intercede mediums of a differing denfity by Prop. 1.

But further, that this difcontinuity of parts is the principal caufe of the opacity of Bodies, will appear by confidering, that opake fubftances become tranfparent by filling their pores with any fubftance of equal or almoft equal denfity with their parts. Thus Paper dipped in Water or Oyl, the *Oculus mundi* Stone fteep'd in Water, Linnen-cloth oyled or varnifhed, and many other fubftances foaked in fuch Liquors as will intimately pervade their little pores, become by that means more tranfparent than otherwife ; fo, on the contrary, the moft tranfparent fubftances may by evacuating their pores, or feparating their parts, be rendred fufficiently opake, as Salts or wet Paper, or the *Oculus mundi* Stone by being dried, Horn by being fcraped, Glafs by being reduced to powder, or otherwife flawed, Turpentine by being ftirred about with Water till they mix imperfectly, and Water by being formed into many fmall Bubbles, either alone in the form of froth, or by fhaking it together with Oyl of Turpentine, or with fome other convenient Liquor, with which it will

not

not perfectly incorporate. And to the increase of the opacity of these Bodies it conduces something, that by the 23th Observation the reflexions of very thin transparent substances are considerably stronger than those made by the same substances of a greater thickness.

P R O P. IV.

The parts of Bodies and their Interstices must not be less than of some definite bigness, to render them opake and coloured.

For the opakest Bodies, if their parts be subtily divided, (as Metals by being dissolved in acid menstruums, &c.) become perfectly transparent. And you may also remember, that in the eighth Observation there was no sensible reflexion at the superficies of the Object-Glasses where they were very near one another, though they did not absolutely touch. And in the 17th Observation the reflexion of the Water-bubble where it became thinnest was almost insensible, so as to cause very black Spots to appear on the top of the Bubble by the want of reflected Light.

On these grounds I perceive it is that Water, Salt, Glass, Stones, and such like substances, are transparent. For, upon divers considerations, they seem to be as full of pores or interstices between their parts as other Bodies are, but yet their parts and interstices to be too small to cause reflexions in their common surfaces.

P R O P.

PROP. V.

The tranſparent parts of Bodies according to their ſe-
veral ſizes muſt reflect rays of one Colour, and tranſmit
thoſe of another, on the ſame grounds that thin Plates or
Bubbles do reflect or tranſmit thoſe rays. And this I take
to be the ground of all their Colours.

For if a thin'd or plated Body, which being of an
even thickneſs, appears all over of one uniform Colour, ſhould be ſlit into threds, or broken into fragments, of the ſame thickneſs with the plate; I ſee no
reaſon why every thred or fragment ſhould not keep its
Colour, and by conſequence why a heap of thoſe threds
or fragments ſhould not conſtitute a maſs or powder of
the ſame Colour, which the plate exhibited before it
was broken. And the parts of all natural Bodies being
like ſo many fragments of a Plate, muſt on the ſame
grounds exhibit the ſame Colours.

Now that they do ſo, will appear by the affinity of
their properties. The finely coloured Feathers of ſome
Birds, and particularly thoſe of Peacocks Tails, do in
the very ſame part of the Feather appear of ſeveral Colours in ſeveral poſitions of the Eye, after the very ſame
manner that thin Plates were found to do in the 7th
and 19th Obſervations, and therefore ariſe from the
thinneſs of the tranſparent parts of the Feathers; that
is, from the ſlenderneſs of the very fine Hairs, or *Capilla-*
menta, which grow out of the ſides of the groſſer late-
ral branches or fibres of thoſe Feathers. And to the
ſame purpoſe it is, that the Webs of ſome Spiders by
being

being spun very fine have appeared coloured, as some have observed, and that the coloured fibres of some silks by varying the position of the Eye do vary their Colour. Also the Colours of silks, cloths, and other substances, which Water or Oyl can intimately penetrate, become more faint and obscure by being immerged in those liquors, and recover their vigor again by being dried, much after the manner declared of thin Bodies in the 10th and 21th Observations. Leaf-gold, some sorts of painted Glass, the infusion of *Lignum Nephriticum*, and some other substances reflect one Colour, and transmit another, like thin Bodies in the 9th and 20th Observations. And some of those coloured powders which Painters use, may have their Colours a little changed, by being very elaborately and finely ground. Where I see not what can be justly pretended for those changes, besides the breaking of their parts into less parts by that contrition after the same manner that the Colour of a thin Plate is changed by varying its thickness. For which reason also it is that the coloured flowers of Plants and Vegitables by being bruised usually become more transparent than before, or at least in some degree or other change their Colours. Nor is it much less to my purpose, that by mixing divers liquors very odd and remarquable productions and changes of Colours may be effected, of which no cause can be more obvious and rational than that the saline corpuscles of one liquor do variously act upon or unite with the tinging corpuscles of another, so as to make them swell, or shrink (whereby not only their bulk but their density also may be changed) or to divide them into smaller corpuscles, (whereby a coloured liquor may become

come tranfparent) or to make many of them affociate into one clufter, whereby two tranfparent liquors may compofe a coloured one. For we fee how apt thofe faline menftruums are to penetrate and diffolve fubftances to which they are applied, and fome of them to precipitate what others diffolve. In like manner, if we confider the various Phænomena of the Atmofphære, we may obferve, that when Vapors are firft raifed, they hinder not the tranfparency of the Air, being divided into parts too fmall to caufe any reflexion in their fuperficies. But when in order to compofe drops of rain they begin to coalefce and conftitute globules of all intermediate fizes, thofe globules when they become of a convenient fize to reflect fome Colours and tranfmit others, may conftitute Clouds of various Colours according to their fizes. And I fee not what can be rationally conceived in fo tranfparent a fubftance as Water for the production of thefe Colours, befides the various fizes of its fluid and globuler parcels.

P R O P. VI.

The parts of Bodies on which their Colours depend, are denfer than the medium, which pervades their interftices.

This will appear by confidering, that the Colour of a Body depends not only on the rays which are incident perpendicularly on its parts, but on thofe alfo which are incident at all other Angles. And that according to the 7th Obfervation, a very little variation of obliquity will change the reflected Colour where the thin body or fmall particle is rarer than the ambient

medium,

medium, infomuch that fuch a fmall particle will at di-
verfly oblique incidences reflect all forts of Colours, in
fo great a variety that the Colour refulting from them
all, confufedly reflected from a heap of fuch particles,
muft rather be a white or grey than any other Colour,
or at beft it muft be but a very imperfect and dirty Colour. Whereas if the thin body or fmall particle be
much denfer than the ambient medium, the Colours
according to the 19th Obfervation are fo little changed
by the variation of obliquity, that the rays which are
reflected leaft obliquely may predominate over the reft
fo much as to caufe a heap of fuch particles to appear
very intenfly of their Colour.

It conduces alfo fomething to the confirmation of this
Propofition, that, according to the 22th Obfervation,
the Colours exhibited by the denfer thin body within
the rarer, are more brifque than thofe exhibited by the
rarer within the denfer.

PROP. VII.

*The bignefs of the component parts of natural Bodies
may be conjectured by their Colours.*

For fince the parts of thefe Bodies by Prop. 5. do
moft probably exhibit the fame Colours with a Plate of
equal thicknefs, provided they have the fame refractive
denfity ; and fince their parts feem for the moft part to
have much the fame denfity with Water or Glafs, as
by many circumftances is obvious to collect ; to determine the fizes of thofe parts you need only have recourfe
to the precedent Tables, in which the thicknefs of Water or Glafs exhibiting any Colour is expreffed. Thus

if

if it be defired to know the Diameter of a corpufcle, which being of equal denfity with Glafs fhall reflect green of the third order; the number $16\frac{1}{4}$ fhews it to be $\frac{16\frac{1}{4}}{100000}$ parts of an Inch.

The greateft difficulty is here to know of what order the Colour of any Body is. And for this end we muft have recourfe to the 4th and 18th Obfervations, from whence may be collected thefe particulars.

Scarlets, and other *reds*, *oranges* and *yellows*, if they be pure and intenfe are moft probably of the fecond order. Thofe of the firft and third order alfo may be pretty good, only the yellow of the firft order is faint, and the orange and red of the third order have a great mixture of violet and blue.

There may be good *greens* of the fourth order, but the pureft are of the third. And of this order the green of all vegitables feem to be, partly by reafon of the intenfenefs of their Colours, and partly becaufe when they wither fome of them turn to a greenifh yellow, and others to a more perfect yellow or orange, or perhaps to red, paffing firft through all the aforefaid intermediate Colours. Which changes feem to be effected by the exhaling of the moifture which may leave the tinging corpufcles more denfe, and fomething augmented by the accretion of the oyly and earthy part of that moifture. Now the green without doubt is of the fame order with thofe Colours into which it changeth, becaufe the changes are gradual, and thofe Colours, though ufually not very full, yet are often too full and lively to be of the fourth order.

Blues

Blues and *purples* may be either of the second or third order, but the beft are of the third. Thus the Colour of violets feems to be of that order, becaufe their Syrup by acid Liquors turns red, and by urinous and alcalizale turns green. For fince it is of the nature of Acids to diffolve or attenuate, and of Alcalies to precipitate or incraffate, if the purple Colour of the Syrup was of the fecond order, an acid Liquor by attenuating its tinging corpufcles would change it to a red of the firft order, and an Alcaly by incraffating them would change it to a green of the fecond order ; which red and green, efpecially the green, feem too imperfect to be the Colours produced by thefe changes. But if the faid purple be fuppofed of the third order, its change to red of the fecond, and green of the third, may without any inconvenience be allowed.

If there be found any Body of a deeper and lefs reddifh purple than that of the violets, its Colour moft probably is of the fecond order. But yet their being no Body commonly known whofe Colour is conftantly more deep than theirs, I have made ufe of their name to denote the deepeft and leaft reddifh purples, fuch as manifeftly tranfcend their Colour in purity.

The *blue* of the firft order, though very faint and little, may poffibly be the Colour of fome fubftances ; and particularly the azure Colour of the Skys feems to be of this order. For all vapours when they begin to condenfe and coalefce into fmall parcels, become firft of that bignefs whereby fuch an Azure muft be reflected before they can conftitute Clouds of other Colours. And fo this being the firft Colour which vapors begin to reflect, it ought to be the Colour of the fineft and moft

tranf-

tranfparent Skys in which vapors are not arrived to that grofnefs requifite to reflect other Colours, as we find it is by experience.

Whitenefs, if moft intenfe and luminous, is that of the firft order, if lefs ftrong and luminous a mixture of the Colours of feveral orders. Of this laft kind is the whitenefs of Froth, Paper, Linnen, and moft white fub- ftances ; of the former I reckon that of white metals to be. For whilft the denfeft of metals, Gold, if foliated is tranfparent, and all metals become tranfparent if diffolved in menftruums or vitrified, the opacity of white metals arifeth not from their denfity alone. They being lefs denfe than Gold would be more tranfparent than it, did not fome other caufe concur with their den- fity to make them opake. And this caufe I take to be fuch a bignefs of their particles as fits them to reflect the white of the firft order. For if they be of other thickneffes they may reflect other Colours, as is mani- feft by the Colours which appear upon hot Steel in tem- pering it, and fometimes upon the furface of melted metals in the Skin or Scoria which arifes upon them in their cooling. And as the white of the firft order is the ftrongeft which can be made by Plates of tranfparent fubftances, fo it ought to be ftronger in the denfer fub- ftances of metals than in the rarer of Air, Water and Glafs. Nor do I fee but that metallic fubftances of fuch a thicknefs as may fit them to reflect the white of the firft order, may, by reafon of their great denfity (accor- ding to the tenour of the firft of thefe Propofitions) re- flect all the Light incident upon them, and fo be as opake and fplendent as its poffible for any Body to be. Gold, or Copper mixed with lefs than half their weight

of

of Silver, or Tin, or Regulus of Antimony, in fufion
or amalgamed with a very little Mercury become white;
which fhews both that the particles of white metals
have much more fuperficies, and fo are fmaller, than
thofe of Gold and Copper, and alfo that they are fo
opake as not to fuffer the particles of Gold or Copper to
fhine through them. Now it is fcarce to be doubted,
but that the Colours of Gold and Copper are of the fecond or third order, and therefore the particles of white
metals cannot be much bigger than is requifite to make
them reflect the white of the firft order. The volati-
lity of Mercury argues that they are not much bigger,
nor may they be much lefs, leaft they lofe their opacity,
and become either tranfparent as they do when attenua-
ted by vitrification, or by folution in menftruums, or
black as they do when ground fmaller, by rubbing Silver, or Tin, or Lead, upon other fubftances to draw black
Lines. The firft and only Colour which white metals
take by grinding their particles fmaller is black, and
therefore their white ought to be that which borders
upon the black Spot in the center of the Rings of Colours, that is, the white of the firft order. But if you
would hence gather the bignefs of metallic particles,
you muft allow for their denfity. For were Mercury
tranfparent, its denfity is fuch that the Sine of incidence upon it (by my computation) would be to the
fine of its refraction, as 71 to 20, or 7 to 2. And
therefore the thicknefs of its particles, that they may
exhibit the fame Colours with thofe of Bubbles of Water, ought to be lefs than the thicknefs of the Skin of
thofe Bubbles in the proportion of 2 to 7. Whence
its poffible that the particles of Mercury may be as little

as

as the particles of fome tranfparent and volatile fluids, and yet reflect the white of the firft order.

Laftly, for the production of *black*, the corpufcles muft be lefs than any of thofe which exhibit Colours. For at all greater fizes there is too much Light reflected to conftitute this Colour. But if they be fuppofed a little lefs than is requifite to reflect the white and very faint blue of the firft order, they will, according to the 4th, 8th, 17th and 18th Obfervations, reflect fo very little as to appear intenfly black, and yet may perhaps varioufly refract it to and fro within themfelves fo long, until it happen to be ftifled and loft, by which means they will appear black in all pofitions of the Eye without any tranfparency. And from hence may be underftood why Fire, and the more fubtile diffolver Putrefaction, by dividing the particles of fubftances, turn them to black, why fmall quantities of black fubftances impart their Colour very freely and intenfly to other fubftances to which they are applied; the minute particles of thefe, by reafon of their very great number, eafily overfpreading the grofs particles of others; why Glafs ground very elaborately with Sand on a copper Plate, 'till it be well polifhed, makes the Sand, together with what is worn off from the Glafs and Copper, become very black: why black fubftances do fooneft of all others become hot in the Sun's Light and burn, (which effect may proceed partly from the multitude of refractions in a little room, and partly from the eafy commotion of fo very fmall corpufcles;) and why blacks are ufually a little inclined to a bluifh Colour. For that they are fo may be feen by illuminating white Paper by Light reflected from black fubftances.

ftances. For the Paper will ufually appear of a bluifh white; and the reafon is, that black borders on the obfcure blue of the firft order defcribed in the 18th Obfervation, and therefore reflects more rays of that Colour than of any other.

In thefe Defcriptions I have been the more particular, becaufe it is not impoffible but that Mifcrofcopes may at length be improved to the difcovery of the particles of Bodies on which their Colours depend, if they are not already in fome meafure arrived to that degree of perfection. For if thofe Inftruments are or can be fo far improved as with fufficient diftinctnefs to reprefent Objects five or fix hundred times bigger than at a Foot diftance they appear to our naked Eyes, I fhould hope that we might be able to difcover fome of the greateft of thofe corpufcles. And by one that would magnify three or four thoufand times perhaps they might all be difcovered, but thofe which produce blacknefs. In the mean while I fee nothing material in this Difcourfe that may rationally be doubted of excepting this Pofition, That tranfparent corpufcles of the fame thicknefs and denfity with a Plate, do exhibit the fame Colour. And this I would have underftood not without fome latitude, as well becaufe thofe corpufcles may be of irregular Figures, and many rays muft be obliquely incident on them, and fo have a fhorter way through them than the length of their Diameters, as becaufe the ftraitnefs of the medium pent in on all fides within fuch corpufcles may a little alter its motions or other qualities on which the reflexion depends. But yet I cannot much fufpect the laft, becaufe I have obferved of fome fmall Plates of Mufcovy-Glafs which were of an

even

even thicknefs, that through a Mifcrofcope they have appeared of the fame Colour at their edges and corners where the included medium was terminated, which they appeared of in other places. However it will add much to our fatisfaction, if thofe corpufcles could be difcovered with Mifcrofcopes ; which if we fhall at length attain to, I fear it will be the utmoft improvement of this fenfe. For it feems impoffible to fee the more fecret and noble works of nature within the corpufcles by reafon of their tranfparency.

PROP. VIII.

The caufe of Reflexion is not the impinging of Light on the folid or impervious parts of Bodies, as is commonly believed.

This will appear by the following Confiderations. Firft, That in the paffage of Light out of Glafs into Air there is a reflexion as ftrong as in its paffage out of Air into Glafs, or rather a little ftronger, and by many degrees ftronger than in its paffage out of Glafs into Water. And it feems not probable that Air fhould have more reflecting parts than Water or Glafs. But if that fhould poffibly be fuppofed, yet it will avail nothing ; for the reflexion is as ftrong or ftronger when the Air is drawn away from the Glafs, (fuppofe in the Air-pump invented by Mr. *Boyle*) as when it is adjacent to it. Secondly, If Light in its paffage out of Glafs into Air be incident more obliquely than at an Angle of 40 or 41 degrees it is wholly reflected, if lefs obliquely it is in great meafure tranfmitted. Now it is not to be imagined that Light at one degree of obliquity fhould meet

K k   with

with pores enough in the Air to tranfmit the greater part of it, and at another degree of obliquity fhould meet with nothing but parts to reflect it wholly, efpecially confidering that in its paffage out of Air into Glafs, how oblique foever be its incidence, it finds pores enough in the Glafs to tranfmit the greateft part of it. If any Man fuppofe that it is not reflected by the Air, but by the outmoft fuperficial parts of the Glafs, there is ftill the fame difficulty : Befides, that fuch a Suppofition is unintelligible, and will alfo appear to be falfe by applying Water behind fome part of the Glafs inftead of Air. For fo in a convenient obliquity of the rays fuppofe of 45 or 46 degrees, at which they are all reflected where the Air is adjacent to the Glafs, they fhall be in great meafure tranfmitted where the Water is adjacent to it; which argues, that their reflexion or tranfmiffion depends on the conftitution of the Air and Water behind the Glafs, and not on the ftriking off the rays upon the parts of the Glafs. Thirdly, If the Colours made by a Prifm placed at the entrance of a beam of Light into a darkened room be fucceffively caft on a fecond Prifm placed at a greater diftance from the former, in fuch manner that they are all alike incident upon it, the fecond Prifm may be fo inclined to the incident rays, that thofe which are of a blue Colour fhall be all reflected by it, and yet thofe of a red Colour pretty copioufly tranfmitted. Now if the reflexion be caufed by the parts of Air or Glafs, I would ask, why at the fame obliquity of incidence the blue fhould wholly impinge on thofe parts fo as to be all reflected, and yet the red find pores enough to be in great meafure tranfmitted. Fourthly, where two Glaffes touch one another,

another, there is no fenfible reflexion as was declared
in the firft Obfervation ; and yet I fee no reafon why
the rays fhould not impinge on the parts of Glafs as
much when contiguous to other Glafs as when contiguous to Air. Fifthly, When the top of a Water-
bubble (in the 17th Obfervation) by the continual fub-
fiding and exhaling of the Water grew very thin, there
was fuch a little and almoft infenfible quantity of Light
reflected from it, that it appeared intenfly black ; whereas round about that black Spot, where the Water was
thicker, the reflexion was fo ftrong as to make the
Water feem very white. Nor is it only at the leaft
thicknefs of thin Plates or Bubbles, that there is no
manifeft reflexion, but at many other thickneffes continually greater and greater. For in the 15th Obfervation the rays of the fame Colour were by turns tranfmitted at one thicknefs, and reflected at another thicknefs, for an indeterminate number of fucceffions. And
yet in the fuperficies of the thinned Body, where it is
of any one thicknefs, there are as many parts for the
rays to impinge on, as where it is of any other thicknefs. Sixthly, If reflexion were caufed by the parts of
reflecting Bodies, it would be impoffible for thin Plates
or Bubbles at the fame place to reflect the rays of one
Colour and tranfmit thofe of another, as they do according to the 13th and 15th Obfervations. For it is
not to be imagined that at one place the rays which
for inftance exhibit a blue Colour, fhould have the for-
tune to dafh upon the parts, and thofe which exhibit
a red to hit upon the pores of the Body ; and then at
another place, where the Body is either a little thicker,
or a little thinner, that on the contrary the blue fhould
K k 2   hit

hit upon its pores, and the red upon its parts. Laftly, were the rays of Light reflected by impinging on the folid parts of Bodies, their reflexions from polifhed Bodies could not be fo regular as they are. For in polifhing Glafs with Sand, Putty or Tripoly, it is not to be imagined that thofe fubftances can by grating and fretting the Glafs bring all its leaft particles to an accurate polifh ; fo that all their furfaces fhall be truly plain or truly fpherical, and look all the fame way, fo as together to compofe one even furface. The fmaller the particles of thofe fubftances are, the fmaller will be the fcratches by which they continually fret and wear away the Glafs until it be polifhed, but be they never fo fmall they can wear away the Glafs no otherwife than by grating and fcratching it, and breaking the proturberances , and therefore polifh it no otherwife than by bringing its roughnefs to a very fine Grain, fo that the fcratches and frettings of the furface become too fmall to be vifible. And therefore if Light were reflected by impinging upon the folid parts of the Glafs, it would be fcattered as much by the moft polifhed Glafs as by the rougheft. So then it remains a Problem, how Glafs polifhed by fretting fubftances can reflect Light fo regularly as it does. And this Problem is fcarce otherwife to be folved than by faying, that the reflexion of a ray is effected, not by a fingle point of the reflecting Body, but by fome power of the Body which is evenly diffufed all over its furface, and by which it acts upon the ray without immediate contact. For that the parts of Bodies do act upon Light at a diftance fhall be fhewn hereafter.

Now

Now if Light be reflected not by impinging on the solid parts of Bodies, but by some other principle ; its probable that as many of its rays as impinge on the solid parts of Bodies are not reflected but stifled and lost in the Bodies. For otherwise we must allow two sorts of reflexions. Should all the rays be reflected which impinge on the internal parts of clear Water or Cryftal, those substances would rather have a cloudy Colour than a clear transparency. To make Bodies look black, its necessary that many rays be stopt, retained and lost in them, and it seems not probable that any rays can be stopt and stifled in them which do not impinge on their parts.

And hence we may understand that Bodies are much more rare and porous than is commonly believed. Water is 19 times lighter, and by consequence 19 times rarer than Gold, and Gold is so rare as very readily and without the least opposition to transmit the magnetick Effluvia, and easily to admit Quick-silver into its pores, and to let Water pass through it. For a concave Sphere of Gold filled with Water, and sodered up, has upon pressing the Sphere with great force, let the Water squeeze through it, and stand all over its outside in multitudes of small Drops, like dew, without bursting or cracking the Body of the Gold as I have been informed by an Eye-witness. From all which we may conclude, that Gold has more pores than solid parts, and by consequence that Water has above forty-times more pores than parts. And he that shall find out an Hypothesis, by which Water may be so rare, and yet not be capable of compression by force, may doubtless by the same Hypothesis make Gold and Water, and all

other

other Bodies as much rarer as he pleaſes, ſo that Light may find a ready paſſage through tranſparent ſub-ſtances.

PROP. IX.

Bodies reflect and refract Light by one and the ſame power variouſly exerciſed in various circumſtances.

This appears by ſeveral Conſiderations. Firſt, Be-cauſe when Light goes out of Glaſs into Air, as ob-liquely as it can poſſibly do, if its incidence be made ſtill more oblique, it becomes totally reflected. For the power of the Glaſs after it has refracted the Light as obliquely as is poſſible if the incidence be ſtill made more oblique, becomes too ſtrong to let any of its rays go through, and by conſequence cauſes total reflexions. Secondly, Becauſe Light is alternately reflected and tranſmitted by thin Plates of Glaſs for many ſucceſſions accordingly, as the thickneſs of the Plate increaſes in an arithmetical Progreſſion. For here the thickneſs of the Glaſs determines whether that power by which Glaſs acts upon Light ſhall cauſe it to be reflected, or ſuffer it to be tranſmitted. And, Thirdly, becauſe thoſe ſurfaces of tranſparent Bodies which have the greateſt refracting power, reflect the greateſt quantity of Light, as was ſhewed in the firſt Propoſition.

PROP. X.

If Light be ſwifter in Bodies than in Vacuo in the proportion of the Sines which meaſure the refraction of the Bodies, the forces of the Bodies to reflect and refract Light,

are

are very nearly proportional to the denfities of the fame Bodies, excepting that unctuous and fulphureous Bodies re- fract more than others of this fame denfity.

Let A B reprefent the refracting plane furface of any Body, and I C a ray incident very obliquely upon the

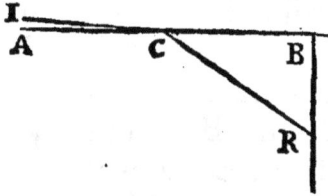

Body in C, fo that the Angle ACI may be infinitely little, and let CR be the refracted ray. From a given point B perpendicular to the refracting furface erect BR meeting with the refracted ray CR in R, and if CR reprefent the motion of the refracted ray, and this motion be diftinguifhed into two motions CB and BR, whereof CB is a parallel to the refracting plane, and BR perpendicular to it : CB fhall reprefent the motion of the incident ray, and BR the motion generated by the refraction, as Opticians have of late explained.

Now if any body or thing in moving through any fpace of a giving breadth terminated on both fides by two parallel plains, be urged forward in all parts of that fpace by forces tending directly forwards towards the laft plain, and before its incidence on the firft plane, had no motion towards it, or but an infinitly little one ; and if the forces in all parts of that fpace, between the planes be at equal diftances from the planes equal to one another, but at feveral diftances be bigger or lefs in any given proportion, the motion generated by the forces in the whole paffage of the body or thing
through

through that space shall be in a subduplicate proportion
of the forces, as Mathematicians will easily understand.
And therefore if the space of activity of the refracting
superficies of the Body be considered as such a space,
the motion of the ray generated by the refracting force
of the Body, during its passage through that space
that is the motion B R must be in a subduplicate
proportion of that refracting force : I say therefore that
the square of the Line B R, and by consequence the
refracting force of the Body is very nearly as the density of the same Body. For this will appear by the following Table, wherein the proportion of the Sines which
measure the refraxions of several Bodies, the square
of B R supposing C B an unite, the densities of the
Bodies estimated by their specifick gravities, and their
refractive power in respect of their densities are set
down in several Columns.

The

The refracting Bodies.	The Proportion of the Sines of incidence and refraction of yellow Light.	The Square of BR, to which the refracting force of the Body is proportionate.	The density and specific gravity of the Body.	The refractive power of the Body in respect of its density.
A Pseudo-Topazius, being a natural, pellucid, brittle, hairy Stone, of a yellow Colour	23 to 14	1'699	4'27	3979
Air	3851 to 3850	0'00052	0'00125	4160
Glaſs of Antimony	17 to 9	2'568	5'28	4864
A Selenitis	61 to 41	1'213	2'252	5386
Glaſs vulgar	31 to 20	1'4025	2'58	5436
Cryſtal of the Rock	25 to 16	1'445	2'65	5450
Iſland Cryſtal	5 to 3	1'778	2'72	6536
Sal Gemmæ	17 to 11	1'388	2'143	6477
Alume	35 to 24	1'1267	1'714	6570
Borax	22 to 15	1'1511	1'714	6716
Niter	32 to 21	1'345	1'9	7079
Dantzick Vitriol	303 to 200	1'295	1'715	7551
Oyl of Vitriol	10 to 7	1'041	1'7	6124
Rain Water	529 to 396	0'7845	1.	7845
Gumm Arabic	31 to 21	1'179	1'375	8574
Spirit of Wine well rectified	100 to 73	0'8765	0'866	10121
Camphire	3 to 2	1'25	0'996	12551
Oyl Olive	22 to 15	1'1511	0'913	12607
Lintſeed Oyl	40 to 27	1'1948	0'932	12819
Spirit of Turpentine	25 to 17	1'1626	0'874	13222
Ambar	14 to 9	1'42	1'04	13654
A Diamond	100 to 41	4'949	3'4	14556

The refraction of the Air in this Table is determined by that of the Atmoſphere obſerved by Aſtronomers. For if Light paſs through many refracting ſubſtances or mediums gradually denſer and denſer, and terminated

with

with parallel furfaces, the fumm of all the refractions will be equal to the fingle refraction which it would have fuffered in paffing immediately out of the firft medium into the laft. And this holds true, though the number of the refracting fubftances be increafed to infinity, and the diftances from one another as much decreafed, fo that the Light may be refracted in every point of its paffage, and by continual refractions bent into a curve Line. And therefore the whole refraction of Light in paffing through the Atmofphere from the higheft and rareft part thereof down to the loweft and denfeft part, muft be equal to the refraction which it would fuffer in paffing at like obliquity out of a Vacuum immediately into Air of equal denfity with that in the loweft part of the Atmofphere.

Now, by this Table, the refractions of a Pfeudo-Topaz, a Selenitis, Rock Cryftal, Ifland Cryftal, Vulgar Glafs (that is, Sand melted together) and Glafs of Antimony, which are terreftrial ftony alcalizate concretes,and Air which probably arifes from fuch fubftances by fermentation,though thefe be fubftances very differing from one another in denfity, yet they have their refractive powers almoft in the fame proportion to one another as their denfities are, excepting that the refraction of that ftrange fubftance Ifland-Cryftal is a little bigger than the reft. And particularly Air, which is 3400 times rarer than the Pfeudo-Topaz, and 4200 times rarer than Glafs of Antimony, has notwithftanding its rarity the fame refractive power in refpect of its denfity which thofe two very denfe fubftances have in refpect of theirs, excepting fo far as thofe two differ from one another.

Again,

Again, the refraction of Camphire, Oyl-Olive, Lint-feed Oyl, Spirit of Turpentine and Amber, which are fat fulphureous unctuous Bodies, and a Diamond, which probably is an unctuous fubftance coagulated, have their refractive powers in proportion to one another as their denfities without any confiderable variation. But the refractive power of thefe unctuous fubftances is two or three times greater in refpect of their denfities than the refractive powers of the former fubftances in refpect of theirs.

Water has a refractive power in a middle degree between thofe two forts of fubftances, and probably is of a middle nature. For out of it grow all vegetable and animal fubftances, which confift as well of fulphureous fat and inflamable parts, as of earthy lean and alcalizate ones.

Salts and Vitriols have refractive powers in a middle degree between thofe of earthy fubftances and Water, and accordingly are compofed of thofe two forts of fubftances. For by diftillation and rectification of their Spirits a great part of them goes into Water, and a great part remains behind in the form of a dry fixt earth capable of vitrification.

Spirit of Wine has a refractive power in a middle degree between thofe of Water and oyly fubftances, and accordingly feems to be compofed of both, united by fermentation ; the Water, by means of fome faline Spirits with which 'tis impregnated, diffolving the Oyl, and volatizing it by the action. For Spirit of Wine is inflamable by means of its oyly parts, and being diftilled often from Salt of Tartar, grows by every diftillation more and more aqueous and flegmatick. And

Chymifts

Chymifts obferve, that Vegitables (as Lavender, Rue, Marjoram, &c.) diftilled *per fe*, before fermentation yield Oyls without any burning Spirits, but after fermentation yield ardent Spirits without Oyls : Which fhews, that their Oyl is by fermentation converted into Spirit. They find alfo, that if Oyls be poured in fmall quantity upon fermentating Vegetables, they diftil over after fermentation in the form of Spirits.

So then, by the foregoing Table, all Bodies feem to have their refractive powers proportional to their denfities, (or very nearly;) excepting fo far as they partake more or lefs of fulphurous oyly particles, and thereby have their refractive power made greater or lefs. Whence it feems rational to attribute the refractive power of all Bodies chiefly, if not wholly, to the fulphurous parts with which they abound. For it's probable that all Bodies abound more or lefs with Sulphurs. And as Light congregated by a Burning-glafs acts moft upon fulphurous Bodies, to turn them into fire and flame; fo, fince all action is mutual, Sulphurs ought to act moft upon Light. For that the action between Light and Bodies is mutual, may appear from this Confideration, That the denfeft Bodies which refract and reflect Light moft ftrongly grow hotteft in the Summer-Sun, by the action of the refracted or reflected Light.

I have hitherto explained the power of Bodies to reflect and refract, and fhewed, that thin tranfparent plates, fibres and particles do, according to their feveral thickneffes and denfities, reflect feveral forts of rays, and thereby appear of feveral Colours, and by confequence that nothing more is requifite for producing all

the

the Colours of natural Bodies than the feveral fizes and denfities of their tranfparent particles. But whence it is that thefe plates, fibres and particles do, according to their feveral thickneffes and denfities, refleft feveral forts of rays, I have not yet explained. To give fome infight into this matter, and make way for underftanding the next Part of this Book, I fhall conclude this Part with a few more Propofitions. Thofe which preceded refpeft the nature of Bodies, thefe the nature of Light : For both muft be underftood before the reafon of their aftions upon one another can be known. And becaufe the laft Propofition depended upon the velocity of Light, I will begin with a Propofition of that kind.

PROP. XI.

Light is propagated from luminous Bodies in time, and fpends about feven or eight minutes of an hour in paffing from the Sun to the Earth.

This was obferved firft by *Romer*, and then by others, by means of the Eclipfes of the Satellites of *Jupiter*. For thefe Eclipfes, when the Earth is between the Sun and *Jupiter*, happen about feven or eight minutes fooner than they ought to do by the Tables, and when the Earth is beyond the Sun they happen about feven or eight minutes later than they ought to do; the reafon being, that the Light of the Satellites has farther to go in the latter cafe than in the former by the Diameter of the Earth's Orbit. Some inequalities of time may arife from the excentricities of the Orbs of the Satellites; but thofe cannot anfwer in all the Satellites, and at all times

to

to the pofition and diftance of the Earth from the Sun.
The mean motions of *Jupiter*'s Satellites is alfo fwifter
in his defcent from his Aphelium to his Perihelium,
than in his afcent in the other half of his Orb : But this
inequality has no refpect to the pofition of the Earth,
and in the three interior Satellites is infenfible, as I find
by computation from the Theory of their gravity.

P R O P. XII.

*Every ray of Light in its paffage through any refra-
cting furface is put into a certain tranfient conftitution
or ftate, which in the progrefs of the ray returns at
equal intervals, and difpofes the ray at every return
to be eafily tranfmitted through the next refracting furface, and between the returns to be eafily reflected by
it.*

This is manifeft by the 5th, 9th, 12th and 15th Obfervations. For by thofe Obfervations it appears, that
one and the fame fort of rays at equal Angles of incidence on any thin tranfparent plate, is alternately reflected and tranfmitted for many fucceffions accordingly,
as the thicknefs of the plate increafes in arithmetical
progreffion of the numbers 0, 1, 2, 3, 4, 5, 6, 7, 8, &c.
fo that if the firft reflexion (that which makes the firft
or innermoft of the Rings of Colours there defcribed)
be made at the thicknefs 1, the rays fhall be tranfmitted at
the thickneffes 0, 2, 4, 6, 8, 10, 12, &c. and thereby
make the central Spot and Rings of Light, which appear by tranfmiffion, and be reflected at the thicknefs
1, 3, 5, 7, 9, 11, &c. and thereby make the Rings which
appear

appear by reflexion. And this alternate reflexion and tranfmiffion, as I gather by the 24th Obfervation, continues for above an hundred viciffitudes, and by the the Obfervations in the next part of this Book, for many thoufands, being propagated from one furface of a Glafsplate to the other, though the thicknefs of the plate be a quarter of an Inch or above : So that this alternation feems to be propagated from every refracting furface to all diftances without end or limitation.

This alternate reflexion and refraction depends on both the furfaces of every thin plate, becaufe it depends on their diftance. By the 21th Obfervation, if either furface of a thin plate of Mufcovy-Glafs be wetted, the Colours caufed by the alternate reflexion and refraction grow faint, and therefore it depends on them both.

It is therefore performed at the fecond furface, for if it were performed at the firft, before the rays arrive at the fecond, it would not depend on the fecond.

It is alfo influenced by fome action or difpofition, propagated from the firft to the fecond, becaufe otherwife at the fecond it would not depend on the firft. And this action or difpofition, in its propagation, intermits and returns by equal intervals, becaufe in all its progrefs it inclines the ray at one diftance from the firft furface to be reflected by the fecond, at another to be tranfmitted by it, and that by equal intervals for innumerable viciffitudes. And becaufe the ray is difpofed to reflexion at the diftances 1, 3, 5, 7, 9, &c. and to tranfmiffion at the diftances 0, 2, 4, 6, 8, 10, &c, (for its tranfmiffion through the firft furface, is at the di-

ftance

ſtance o, and it is tranſmitted through both together, if their diſtance be infinitely little or much leſs
than 1) the diſpoſition to be tranſmitted at the diſtances
2, 4, 6, 8, 10, &c. is to be accounted a return of the
ſame diſpoſition which the ray firſt had at the diſtance o,
that is at its tranſmiſſion through the firſt refracting ſurface. All which is the thing I would prove.

What kind of action or diſpoſition this is? Whether
it conſiſt in a circulating or a vibrating motion of the
ray, or of the medium, or ſomething elſe? I do not
here enquire. Thoſe that are averſe from aſſenting to
any new diſcoveries, but ſuch as they can explain by an
Hypotheſis, may for the preſent ſuppoſe, that as Stones
by falling upon Water put the Water into an undula-
ting motion, and all Bodies by percuſſion excite vibrations in the Air; ſo the rays of Light, by impinging on
any refracting or reflecting ſurface, excite vibrations in
the refracting or reflecting medium or ſubſtance, and
by exciting them agitate the ſolid parts of the refracting
or reflecting Body, and by agitating them cauſe the Body
to grow warm or hot; that the vibrations thus excited
are propagated in the refracting or reflecting medium
or ſubſtance, much after the manner that vibrations are
propagated in the Air for cauſing ſound, and move
faſter than the rays ſo as to overtake them; and that
when any ray is in that part of the vibration which con-
ſpires with its motion, it eaſily breaks through a refracting ſurface, but when it is in the contrary part of
the vibration which impedes its motion, it is eaſily
reflected; and, by conſequence, that every ray is ſucceſſively diſpoſed to be eaſily reflected, or eaſily tranſmitted, by every vibration which overtakes it. But
whether

whether tnis Hypotnefis be true or falfe I do not here
confider. I content my felf with the bare difcovery,
that the rays of Light are by fome caufe or other alternately difpofed to be reflected or refracted for many viciffitudes.

DEFINITION.

*The returns of the difpofition of any ray to be reflected
I will call its* Fits of eafy reflexion, *and thofe of
its difpofition to be tranfmitted its* Fits of eafy tranfmiffion, *and the fpace it paffes between every return and the next return*, *the* Interval of its
Fits.

PROP. XIII.

*The reafon why the furfaces of all thick tranfparent
Bodies reflect part of the Light incident on them, and
refract the reft, is, that fome rays at their incidence are
in Fits of eafy reflexion, and others in Fits of eafy tranfmiffion.*

This may be gathered from the 24th Obfervation,
where the Light reflected by thin plates of Air and Glafs,
which to the naked Eye appeared evenly white all over
the plate, did through a Prifm appear waved with many
fucceffions of Light and Darknefs made by alternate fits
of eafy reflexion and eafy tranfmiffion, the Prifm
fevering and diftinguifhing the waves of which the
white reflected Light was compofed, as was explained
above.

M m   And

And hence Light is in fits of easy reflexion and easy transmission, before its incidence on transparent Bodies. And probably it is put into such fits at its first emission from luminous Bodies, and continues in them during all its progress. For these fits are of a lasting Nature, as will appear by the next part of this Book.

In this Proposition I suppose the transparent Bodies to be thick, because if the thickness of the Body be much less than the interval of the fits of easy reflexion and transmission of the rays, the Body loseth its reflecting power. For if the rays, which at their entering into the Body are put into fits of easy transmission, arrive at the furthest surface of the Body before they be out of those fits they must be transmitted. And this is the reason why Bubbles of Water lose their reflecting power when they grow very thin, and why all opake Bodies when reduced into very small parts become transparent.

PROP. XIV.

Those surfaces of transparent Bodies, which if the ray be in a fit of refraction do refract it most strongly, if the ray be in a fit of reflexion do reflect it most easily.

For we shewed above in Prop. 8. that the cause of reflexion is not the impinging of Light on the solid impervious parts of Bodies, but some other power by which those solid parts act on Light at a distance. We shewed also in Prop. 9. that Bodies reflect and refract Light by one and the same power variously exercised in various circumstances, and in Prop. 1. that the most strongly refracting surfaces reflect the most Light: All
which

which compared together evince and ratify both this and the laſt Propoſition.

P R O P. XV.

In any one and the ſame ſort of rays emerging in any Angle out of any refracting ſurface into one and the ſame medium, the interval of the following fits of eaſy reflexion and tranſmiſſion are either accurately or very nearly, as the Rectangle of the ſecant of the Angle of refraction, and of the ſecant of another Angle, whoſe ſine is the firſt of 106 arithmetical mean proportionals, between the ſines of incidence and refraction counted from the ſine of refraction.

This is manifeſt by the 7th Obſervation.

P R O P. XVI.

In ſeveral ſorts of rays emerging in equal Angles out of any refracting ſurface into the ſame medium, the intervals of the following fits of eaſy reflexion and eaſy tranſmiſſion are either accurately, or very nearly, as the Cube-roots of the Squares of the lengths of a Chord, which found the notes in an Eight, ſol, la, fa, ſol, la, mi, fa, ſol, with all their intermediate degrees anſwering to the Colours of thoſe rays, according to the Analogy deſcribed in the ſeventh Experiment of the ſecond Book.

This is manifeſt by the 13th and 14th Obſervations.

PROP.

P R O P. XVII.

*If rays of any one sort pass perpendicularly into several
mediums, the intervals of the fits of easy reflexion and
transmission in any one medium, is to those intervals in
any other as the sine of incidence to the sine of refraction,
when the rays pass out of the first of those two mediums
into the second.*

This is manifest by the 10th Observation.

P R O P. XVIII.

*If the rays which paint the Colour in the confine of
yellow and orange pass perpendicularly out of any medium
into Air, the intervals of their fits of easy reflexion are
the $\frac{1}{89000}$th part of an Inch. And of the same length are
the intervals of their fits of easy transmission.*

This is manifest by the 6th Observation.

From these Propositions it is easy to collect the intervals of the fits of easy reflexion and easy transmission of any sort of rays refracted in any Angle into
any medium, and thence to know, whether the rays
shall be reflected or transmitted at their subsequent
incidence upon any other pellucid medium. Which
thing being useful for understanding, the next part of
this Book was here to be set down. And for the same
reason I add the two following Propositions.

P R O P.

P R O P. XIX.

If any fort of rays falling on the polite furface of any pellucid medium be reflected back, the fits of eafy reflexion which they have at the point of reflexion, fhall ftill continue to return, and the returns fhall be at diftances from the point of reflexion in the arithmetical progreffion of the numbers 2, 4, 6, 8, 10, 12, &c. *and between thefe fits the rays fhall be in fits of eafy tranfmiffion.*

For fince the fits of eafy reflexion and eafy tranfmiffion are of a returning nature, there is no reafon why thefe fits, which continued till the ray arrived at the reflecting medium, and there inclined the ray to reflexion, fhould there ceafe. And if the ray at the point of reflexion was in a fit of eafy reflexion, the progreffion of the diftances of thefe fits from that point muft begin from 0, and fo be of the numbers 0, 2, 4, 6, 8, &c. And therefore the progreffion of the diftances of the intermediate fits of eafy tranfmiffion reckoned from the fame point, muft be in the progreffion of the odd numbers 1, 3, 5, 7, 9, &c. contrary to what happens when the fits are propagated from points of refraction.

P R O P. XX.

The intervals of the fits of eafy reflexion and eafy tranfmiffion, propagated from points of reflexion into any medium, are equal to the intervals of the like fits which the fame rays would have, if refracted into the fame medium

medium in Angles of refraction equal to their Angles of reflexion.

For when Light is reflected by the second surface of thin plates, it goes out afterwards freely at the first surface to make the Rings of Colours which appear by reflexion, and by the freedom of its egress, makes the Colours of these Rings more vivid and strong than those which appear on the other side of the plates by the transmitted Light. The reflected rays are therefore in fits of easy transmission at their egress ; which would not always happen, if the intervals of the fits within the plate after reflexion were not equal both in length and number to their intervals before it. And this confirms also the proportions set down in the former Proposition. For if the rays both in going in and out at the first surface be in fits of easy transmission, and the intervals and numbers of those fits between the first and second surface, before and after reflexion, be equal ; the distances of the fits of easy transmission from either surface, must be in the same progression after reflexion as before ; that is, from the first surface which transmitted them, in the progression of the even numbers 0, 2, 4, 6, 8, &c. and from the second which reflected them, in that of the odd numbers 1, 3, 5, 7, &c. But these two Propositions will become much more evident by the Observations in the following part of this Book.

T H E

THE

SECOND BOOK

OF

OPTICKS.

PART IV.

Obfervations concerning the Reflexions and Colours of thick tranfparent polifhed Plates.

THere is no Glafs or Speculum how well foever polifhed, but, befides the Light which it refracts or reflects regularly, fcatters every way irregularly a faint Light, by means of which the polifhed furface, when illuminated in a dark Room by a beam of the Sun's Light, may be eafily feen in all pofitions of the Eye. There are certain Phænomena of this fcattered Light, which when I firft obferved them, feemed very ftrange and furprifing to me. My Obfervations were as follows.

OBS.

O B S. I.

The Sun fhining into my darkened Chamber through
a Hole $\frac{1}{3}$ of an Inch wide, I let the intromitted beam
of Light fall perpendicularly upon a Glafs Speculum
ground concave on one fide and convex on the other,
to a Sphere of five Feet and eleven Inches Radius, and
quick-filvered over on the convex fide. And holding
a white opake Chart, or a Quire of Paper at the Center
of the Spheres to which the Speculum was ground, that
is, at the diftance of about five Feet and eleven Inches
from the Speculum, in fuch manner, that the beam of
Light might pafs through a little Hole made in the
middle of the Chart to the Speculum, and thence be
reflected back to the fame Hole : I obferved upon the
Chart four or five concentric Irifes or Rings of Colours,
like Rain-bows, encompaffing the Hole much after the
manner that thofe, which in the fourth and following
Obfervations of the firft part of this third Book appeared
between the Object-Glaffes, encompaffed the black Spot,
but yet larger and fainter than thofe. Thefe Rings as
they grew larger and larger became diluter and fainter,
fo that the fifth was fcarce vifible. Yet fometimes,
when the Sun fhone very clear, there appeared faint
Lineaments of a fixth and feventh. If the diftance of
the Chart from the Speculum was much greater or much
lefs than that of fix Feet, the Rings became dilute and
vanifhed. And if the diftance of the Speculum from
the Window was much greater than that of fix Feet,
the reflected beam of Light would be fo broad at the
diftance of fix Feet from the Speculum where the Rings
appeared,

appeared, as to obfcure one or two of the innermoft
Rings. And therefore I ufually placed the Speculum
at about fix Feet from the Window; fo that its Focus
might there fall in with the center of its concavity at the
Rings upon the Chart. And this pofture is always to
be underftood in the following Obfervations where no
other is expreft.

O B S. II.

The Colours of thefe Rain-bows fucceeded one another from the center outwards, in the fame form and
order with thofe which were made in the ninth Obfervation of the firft Part of this Book by Light not reflected, but tranfmitted through the two Object-Glaffes.
For, firft, there was in their common center a white
round Spot of faint Light, fomething broader than the
reflected beam of Light; which beam fometimes fell
upon the middle of the Spot, and fometimes by a little
inclination of the Speculum receded from the middle,
and left the Spot white to the center.

This white Spot was immediately encompaffed with
a dark grey or ruffet, and that darknefs with the Colours of the firft Iris, which were on the infide next
the darknefs a little violet and indico, and next to that
a blue, which on the outfide grew pale, and then fucceeded a little greenifh yellow, and after that a brighter
yellow, and then on the outward edge of the Iris a red
which on the outfide inclined to purple.

This Iris was immediately encompaffed with a fecond, whofe Colours were in order from the infide

N n   out-

outwards, purple, blue, green, yellow, light red, a red mixed with purple.

Then immediately followed the Colours of the third Iris, which were in order outwards a green inclining to purple, a good green, and a red more bright than that of the former Iris.

The fourth and fifth Iris feemed of a bluifh green within, and red without, but fo faintly that it was difficult to difcern the Colours.

O B S. III.

Meafuring the Diameters of thefe Rings upon the Chart as accurately as I could, I found them alfo in the fame proportion to one another with the Rings made by Light tranfmitted through the two Object-Glaffes. For the Diameters of the four firft of the bright Rings meafured between the brighteft parts of their orbits, at the diftance of fix Feet from the Speculum were $1\frac{11}{16}$, $2\frac{3}{8}$, $2\frac{11}{12}$, $3\frac{3}{8}$ Inches, whofe fquares are in arithmetical progreffion of the numbers 1, 2, 3, 4. If the white circular Spot in the middle be reckoned amongft the Rings, and its central Light, where it feems to be moft luminous, be put equipollent to an infinitely little Ring; the fquares of the Diameters of the Rings will be in the progreffion 0, 1, 2, 3, 4, &c. I meafured alfo the Diameters of the dark Circles between thefe luminous ones, and found their fquares in the progreffion of the numbers $\frac{1}{2}$, $1\frac{1}{2}$, $2\frac{1}{2}$, $3\frac{1}{2}$, &c. the Diameters of the firft four at the diftance of fix Feet from the Speculum, being $1\frac{3}{16}$, $2\frac{1}{16}$, $2\frac{2}{3}$, $3\frac{3}{20}$ Inches. If the diftance of the Chart from the Speculum was increafed

creafed or diminifhed, the Diameters of the Circles were increafed or diminifhed proportionally.

O B S. IV.

By the analogy between thefe Rings and thofe defcribed in the Obfervations of the firft Part of this Book, I fufpeƈted that there were many more of them which fpread into one another, and by interfering mixed their Colours, and diluted one another fo that they could not be feen apart.   I viewed them therefore through a Prifm, as I did thofe in the 24th Obfervation of the firft Part of this Book.   And when the Prifm was fo placed as by refraƈting the Light of their mixed Colours to feparate them, and diftinguifh the Rings from one another, as it did thofe in that Obfervation, I could then fee them diftinƈter than before, and eafily number eight or nine of them, and fometimes twelve or thirteen.   And had not their Light been fo very faint, I queftion not but that I might have feen many more.

O B S. V.

Placing a Prifm at the Window to refraƈt the intromitted beam of Light, and caft the oblong Speƈtrum of Colours on the Speculum : I covered the Speculum with a black Paper which had in the middle of it a Hole to let any one of the Colours pafs through to the Speculum, whilft the reft were intercepted by the Paper. And now I found Rings of that Colour only which fell upon the Speculum.   If the Speculum was illuminated with red the Rings were totally red with dark inter-

N n 2   vals,

vals, if with blue they were totally blue, and fo of the other Colours. And when they were illuminated with any one Colour, the Squares of their Diameters mea-fured between their moft luminous parts, were in the arithmetical progreffion of the numbers 0, 1, 2, 3, 4, and the Squares of the Diameters of their dark intervals in the progreffion of the intermediate numbers $\frac{1}{2}$, $1\frac{1}{2}$, $2\frac{1}{2}$, $3\frac{1}{2}$: But if the Colour was varied they varied their magni-tude. In the red they were largeft, in the indico and violet leaft, and in the intermediate Colours yellow, green and blue; they were of feveral intermediate big-neffes anfwering to the Colour, that is, greater in yel-low than in green, and greater in green than in blue. And hence I knew that when the Speculum was illumi-nated with white Light, the red and yellow on the out-fide of the Rings were produced by the leaft refrangible rays, and the blue and violet by the moft refrangible, and that the Colours of each Ring fpread into the Co-lours of the neighbouring Rings on either fide, after the manner explained in the firft and fecond Part of this Book, and by mixing diluted one another fo that they could not be diftinguifhed, unlefs near the center where they were leaft mixed. For in this Obfervation I could fee the Rings more diftinctly, and to a greater number than before, being able in the yellow Light to number eight or nine of them, befides a faint fhadow of a tenth. To fatisfy my felf how much the Colours of the feveral Rings fpread into one another, I meafured the Diame-ters of the fecond and third Rings, and found them when made by the confine of the red and orange to be the fame Diameters when made by the confine of blue and indico, as 9 to 8, or thereabouts. For it was hard

to

to determine this proportion accurately. Alſo the Circles made ſucceſſively by the red, yellow and green,
differed more from one another than thoſe made ſuccef-
ſively by the green, blue and indico. For the Circle
made by the violet was too dark to be ſeen. To carry
on the computation, Let us therefore ſuppoſe that the
differences of the Diameters of the Circles made by the
outmoſt red, the confine of red and orange, the confine
of orange and yellow, the confine of yellow and green,
the confine of green and blue, the confine of blue and
indico, the confine of indico and violet, and outmoſt violet, are in proportion as the differences of the lengths
of a Monochord which ſound the tones in an Eight ;
ſol, la, fa, ſol, la, mi, fa, ſol, that is, as the numbers $\frac{1}{9}$,
$\frac{1}{18}$, $\frac{1}{12}$, $\frac{1}{12}$, $\frac{1}{27}$, $\frac{1}{27}$, $\frac{1}{18}$. And if the Diameter of the Circle made
by the confine of red and orange be 9 A, and that of
the Circle made by the confine of blue and indico be
8 A as above, their difference 9 A ---- 8 A will be to
the difference of the Diameters of the Circles made by
the outmoſt red, and by the confine of red and orange,
as $\frac{1}{18} + \frac{1}{12} + \frac{1}{12} + \frac{1}{27}$ to $\frac{1}{9}$, that is as $\frac{8}{9}$ to $\frac{1}{3}$ or 8 to 3, and to
the difference of the Circles made by the outmoſt violet,
and by the confine of blue and indico, as $\frac{1}{18} + \frac{1}{12} + \frac{1}{12} + \frac{1}{9}$
to $\frac{1}{27} + \frac{1}{18}$, that is, as $\frac{8}{9}$ to $\frac{1}{4}$, or as 16 to 5. And therefore theſe differences will be $\frac{1}{3}$ A and $\frac{5}{16}$ A. Add the
firſt to 9 A and ſubduct the laſt from 8 A, and you
will have the Diameters of the Circles made by the
leaſt and moſt refrangible rays $\frac{1}{8}$ A and $\frac{61\frac{1}{2}}{8}$ A. Theſe
Diameters are therefore to one another as 75 to 61½ or
50 to 41, and their Squares as 2500 to 1681, that is,
as 3 to 2 very nearly. Which proportion differs not
much from the proportion of the Diameters of the
Circles

Circles made by the outmoſt red and outmoſt violet in the 13th Obſervation of the firſt part of this Book.

O B S. VI.

Placing my Eye where theſe Rings appeared plaineſt, I ſaw the Speculum tinged all over with waves of Colours (red, yellow, green, blue ;) like thoſe which in the Obſervations of the firſt Part of this Book appeared between the Object-Glaſſes and upon Bubbles of Water, but much larger. And after the manner of thoſe, they were of various magnitudes in various poſitions of the Eye, ſwelling and ſhrinking as I moved my Eye this way and that way. They were formed like Arcs of concentrick Circles as thoſe were, and when my Eye was over againſt the center of the concavity of the Speculum (that is, 5 Feet and 10 Inches diſtance from the Speculum) their common center was in a right Line with that center of concavity, and with the Hole in the Window. But in other poſtures of my Eye their center had other poſitions. They appeared by the Light of the Clouds propagated to the Speculum through the Hole in the Window, and when the Sun ſhone through that Hole upon the Speculum, his Light upon it was of the Colour of the Ring whereon it fell, but by its ſplendor obſcured the Rings made by the Light of the Clouds, unleſs when the Speculum was removed to a great diſtance from the Window, ſo that his Light upon it might be broad and faint. By varying the poſition of my Eye, and moving it nearer to or farther from the direct beam of the Sun's Light, the Colour of the Sun's reflected Light conſtantly varied upon the Speculum,

as

as it did upon my Eye, the fame Colour always appearing to a By-ftander upon my Eye which to me appeared upon the Speculum. And thence I knew that
the Rings of Colours upon the Chart were made by thefe
reflected Colours propagated thither from the Speculum in feveral Angles, and that their production depended not upon the termination of Light and Shad-
dow.

OBS. VII.

By the Analogy of all thefe Phænomena with thofe of
the like Rings of Colours defcribed in the firft Part of
this Book, it feemed to me that thefe Colours were
produced by this thick plate of Glafs, much after the
manner that thofe were produced by very thin
plates. For, upon tryal, I found that if the Quick-
filver were rubbed off from the back-fide of the Speculum, the Glafs alone would caufe the fame Rings of
Colours, but much more faint than before ; and therefore the Phænomenon depends not upon the Quick-
filver, unlefs fo far as the Quick-filver by the increafing
the reflexion of the back-fide of the Glafs increafes the
Light of the Rings of Colours. I found alfo that a Speculum of metal without Glafs made fome years fince
for optical ufes, and very well wrought, produced none
of thofe Rings; and thence I underftood that thefe
Rings arife not from one fpecular furface alone, but
depend upon the two furfaces of the plate of Glafs whereof the Speculum was made, and upon the thicknefs of
the Glafs between them. For as in the 7th and 19th
Obfervations of the firft Part of this Book a thin plate
of

of Air, Water, or Glafs of an even thicknefs appeared
of one Colour when the rays were perpendicular to it,
of another when they were a little oblique, of another
when more oblique, of another when ftill more oblique,
and fo on ; fo here, in the fixth Obfervation, the Light
which emerged out of the Glafs in feveral obliquities,
made the Glafs appear of feveral Colours, and being
propagated in thofe obliquities to the Chart, there painted Rings of thofe Colours. And as the reafon why a
thin plate appeared of feveral Colours in feveral obliquities of the rays, was, that the rays of one and the fame
fort are reflected by the thin plate at one obliquity and
tranfmitted at another, and thofe of other forts tranfmitted where thefe are reflected, and reflected where
thefe are tranfmitted : So the reafon why the thick
plate of Glafs whereof the Speculum was made did appear of various Colours in various obliquities, and in
thofe obliquities propagated thofe Colours to the Chart,
was, that the rays of one and the fame fort did at one
obliquity emerge out of the Glafs, at another did not
emerge but were reflected back towards the Quick-filver by the hither furface of the Glafs, and accordingly
as the obliquity became greater and greater emerged
and were reflected alternately for many fucceffions, and
that in one and the fame obliquity the rays of one fort
were reflected, and thofe of another tranfmitted. This
is manifeft by the firft Obfervat.on of this Book : For
in that Obfervation, when the Speculum was illuminated by any one of the prifmatick Colours, that Light
made many Rings of the fame Colour upon the Chart
with dark intervals, and therefore at its emergence out
of the Speculum was alternately tranfmitted, and not
tranf-

tranfmitted from the Speculum to the Chart for many fucceffions, according to the various obliquities of its emergence. And when the Colour caft on the Speculum by the Prifm was varied, the Rings became of the Colour caft on it, and varied their bignefs with their Colour, and therefore the Light was now alternately tranfmitted and not tranfmitted from the Speculum to the Lens at other obliquities than before. It feemed to me therefore that thefe Rings were of one and the fame original with thofe of thin plates, but yet with this difference that thofe of thin plates are made by the alternate reflexions and tranfmiffions of the rays at the fecond furface of the plate after one paffage through it: But here the rays go twice through the plate before they are alternately reflected and tranfmitted; firft, they go through it from the firft furface to the Quickfilver, and then return through it from the Quick-filver to the firft furface, and there are either tranfmitted to the Chart or reflected back to the Quick-filver, accordingly as they are in their fits of eafie reflexion or tranfmiffion when they arrive at that furface. For the intervals of the fits of the rays which fall perpendicularly on the Speculum, and are reflected back in the fame perpendicular Lines, by reafon of the equality of thefe Angles and Lines, are of the fame length and number within the Glafs after reflexion as before by the 19th Propofition of the third Part of this Book. And therefore fince all the rays that enter through the firft furface are in their fits of eafy tranfmiffion at their entrance, and as many of thefe as are reflected by the fecond are in their fits of eafy reflexion there, all thefe muft be again in their fits of eafy tranfmiffion at their

return

return to the firft, and by confequence there go out of the Glafs to the Chart, and form upon it the white Spot of Light in the center of the Rings. For the reafon holds good in all forts of rays, and therefore all forts muft go out promifcuoufly to that Spot, and by their mixture caufe it to be white. But the intervals of the fits of thofe rays which are reflected more obliquely than they enter, muft be greater after reflexion than before by the 15th and 20th Prop. And thence it may happen that the rays at their return to the firft furface, may in certain obliquities be in fits of eafy reflexion, and return back to the Quick-filver, and in other intermediate obliquities be again in fits of eafy tranfmiffion, and fo go out to the Chart, and paint on it the Rings of Colours about the white Spot. And becaufe the intervals of the fits at equal obliquities are greater and fewer in the lefs refrangible rays, and lefs and more numerous in the more refrangible, therefore the lefs refrangible at equal obliquities fhall make fewer Rings than the more refrangible, and the Rings made by thofe fhall be larger than the like number of Rings made by thefe ; that is, the red Rings fhall be larger than the yellow, the yellow than the green, the green than the blue, and the blue than the violet, as they were really found to be in the 5th Obfervation. And therefore the firft Ring of all Colours incompaffing the white Spot of Light fhall be red without and violet within, and yellow, and green, and blue in the middle, as it was found in the fecond Obfervation; and thefe Colours in the fecond Ring, and thofe that follow fhall be more expanded till they fpread into one another, and blend one another by interfering.

Thefe

Thefe feem to be the reafons of thefe Rings in general, and this put me upon obferving the thicknefs of
the Glafs, and confidering whether the dimenfions and
proportions of the Rings may be truly derived from it
by computation.

O B S. VIII.

I meafured therefore the thicknefs of this concavo-
convex plate of Glafs, and found it every-where $\frac{1}{4}$ of an
Inch precifely. Now, by the 6th Obfervation of the
firft Part of this Book, a thin plate of Air tranfmits the
brighteft Light of the firft Ring, that is the bright yellow, when its thicknefs is the $\frac{1}{89000}$th part of an Inch,
and by the 10th Obfervation of the fame part, a thin
plate of Glafs tranfmits the fame Light of the fame Ring
when its thicknefs is lefs in proportion of the fine of
refraction to the fine of incidence, that is, when its
thicknefs is the $\frac{11}{1513000}$th or $\frac{1}{137545}$th part of an Inch, fuppofing the fines are as 11 to 17. And if this thicknefs
be doubled it tranfmits the fame bright Light of the
fecond Ring, if tripled it tranfmits that of the third,
and fo on, the bright yellow Light in all thefe cafes being in its fits of tranfmiffion. And therefore if its thicknefs be multiplied 34386 times fo as to become $\frac{1}{4}$ of an
Inch it tranfmits the fame bright Light of the 34386th
Ring. Suppofe this be the bright yellow Light tranfmitted perpendicularly from the reflecting convex fide
of the Glafs through the concave fide to the white Spot
in the center of the Rings of Colours on the Chart : And
by a rule in the feventh Obfervation in the firft Part of
the firft Book, and by the 15th and 20th Propofitions

O o 2   of

of the third Part of this Book, if the rays be made oblique to the Glaſs, the thickneſs of the Glaſs requiſite to tranſmit the ſame bright Light of the ſame Ring in any obliquity is to this thickneſs of ¼ of an Inch, as the ſecant of an Angle whoſe ſine is the firſt of an hundred and ſix arithmetical means between the ſines of incidence and refraction, counted from the ſine of incidence when the refraction is made out of any plated Body into any medium incompaſſing it, that is, in this caſe, out of Glaſs into Air. Now if the thickneſs of the Glaſs be increaſed by degrees, ſo as to bear to its firſt thickneſs, (*viz.* that of a quarter of an Inch) the proportions which 34386 (the number of fits of the perpendicular rays in going through the Glaſs towards the white Spot in the center of the Rings,) hath to 34385, 34384, 34383 and 34382 (the numbers of theſits of the oblique rays in going through the Glaſs towards the firſt, ſecond, third and fourth Rings of Colours,) and if the firſt thickneſs be divided into 100000000 equal parts, the increaſed thickneſſes will be 100002908, 100005816, 100008725 and 100011633, and the Angles of which theſe thickneſſes are ſecants will be 26′ 13″, 37′ 5″, 45′ 6″ and 52′ 26″, the Radius being 100000000 ; and the ſines of theſe Angles are 762, 1079, 1321 and 1525, and the proportional ſines of refraction 1172, 1659, 2031 and 2345, the Radius being 100000. For ſince the ſines of incidence out of Glaſs into Air are to the ſines of refraction as 11 to 17, and to the above-mentioned ſecants as 11 to the firſt of 106 arithmetical means between 11 and 17, that is as 11 to $11\frac{6}{106}$, thoſe ſecants will be to the ſines of refraction as $11\frac{6}{106}$ to 17, and by this Analogy will give theſe ſines. So then

if

if the obliquities of the rays to the concave furface of
the Glafs be fuch that the fines of their refraction in
paffing out of the Glafs through that furface into the
Air be 1172, 1659, 2031, 2345, the bright Light of
the 34386th Ring fhall emerge at the thickneffes of the
Glafs which are to $\frac{1}{4}$ of an Inch as 34386 to 34385,
34384, 34383, 34382, refpectively. And therefore if
the thicknefs in all thefe cafes be $\frac{1}{4}$ of an Inch (as it is in
the Glafs of which the Speculum was made) the bright
Light of the 34385th Ring fhall emerge where the fine
of refraction is 1172, and that of the 34384th, 384383th
and 34382th Ring where the fine is 1659, 2031, and
2345 refpectively. And in thefe Angles of refraction
the Light of thefe Rings fhall be propagated from the
Speculum to the Chart, and there paint Rings about the
white central round Spot of Light which we faid was
the Light of the 34386th Ring. And the Semidiame-
ters of thefe Rings fhall fubtend the Angles of refraction
made at the concave furface of the Speculum, and by
confequence their Diameters fhall be to the diftance of
the Chart from the Speculum as thofe fines of refraction
doubled are to the Radius that is as 1172, 1659, 2031,
and 2345, doubled are to 100000. And therefore if
the diftance of the Chart from the concave furface of
the Speculum be fix Feet (as it was in the third of thefe
Obfervations) the Diameters of the Rings of this bright
yellow Light upon the Chart fhall be 1'688, 2'389,
2'925, 3'375 Inches : For thefe Diameters are to 6 Feet
as the above-mentioned fines doubled are to the Radius.
Now thefe Diameters of the bright yellow Rings, thus
found by computation are the very fame with thofe
found in the third of thefe Obfervations by meafuring
them,

them, (*viz.* with $1\frac{11}{16}$, $2\frac{3}{8}$, $2\frac{11}{12}$, and $3\frac{3}{8}$ Inches, and therefore the Theory of deriving these Rings from the thickness of the plate of Glass of which the Speculum was made, and from the obliquity of the emerging rays agrees with the Observation. In this computation I have equalled the Diameters of the bright Rings made by Light of all Colours, to the Diameters of the Rings made by the bright yellow. For this yellow makes the brightest part of the Rings of all Colours. If you desire the Diameters of the Rings made by the Light of any other unmixed Colour, you may find them readily by putting them to the Diameters of the bright yellow ones in a subduplicate proportion of the intervals of the fits of the rays of those Colours when equally inclined to the refracting or reflecting surface which caused those fits, that is, by putting the Diameters of the Rings made by the rays in the extremities and limits of the seven Colours, red, orange, yellow, green, blue, indico, violet, proportional the Cube-roots of the numbers, 1, $\frac{8}{9}$, $\frac{5}{6}$, $\frac{3}{4}$, $\frac{2}{3}$, $\frac{3}{5}$, $\frac{9}{16}$, $\frac{1}{2}$, which express the lengths of a Monochard sounding the notes in an Eight : For by this means the Diameter of the Rings of these Colours will be found pretty nearly in the same proportion to one another, which they ought to have by the fifth of these Observations.

And thus I satisfied my self that these Rings were of the same kind and original with those of thin plates, and by consequence that the fits or alternate dispositions of the rays to be reflected and transmitted are propagated to great distances from every reflecting and refracting surface. But yet to put the matter out of doubt I added the following Observation.

O B S.

O B S. IX.

If thefe Rings thus depend on the thicknefs of the plate
of Glafs their Diameters at equal diftances from feveral
Speculums made of fuch concavo-convex plates of Glafs
as are ground on the fame Sphere, ought to be recipro-
cally in a fubduplicate proportion of the thickneffes of
the plates of Glafs. And if this proportion be found
true by experience it will amount to a demonftration
that thefe Rings (like thofe formed in thin plates) do
depend on the thicknefs of the Glafs. I procured therefore another concavo-convex plate of Glafs ground on
both fides to the fame Sphere with the former plate :
Its thicknefs was $\frac{5}{62}$ parts of an Inch ; and the Diameters
of the three firft bright Rings meafured between the
brighteft parts of their orbits at the diftance of 6 Feet
from the Glafs were 3. $4\frac{1}{6}$. $5\frac{1}{8}$. Inches. Now the thicknefs of the other Glafs being $\frac{1}{4}$ of an Inch was to thicknefs of this Glafs as $\frac{1}{4}$ to $\frac{5}{62}$, that is as 31 to 10, or
31000000 to 10000000, and the roots of thefe numbers
are 17607 and 10000, & in the proportion of the firft
of thefe roots to the fecond are the Diameters of the
bright Rings made in this Obfervation by the thinner
Glafs, 3. $4\frac{1}{6}$. $5\frac{1}{8}$ to the Diameters of the fame Rings made
in the third of thefe Obfervations by the thicker Glafs
$1\frac{11}{16}$. $2\frac{3}{8}$ $2\frac{11}{12}$, that is, the Diameters of the Rings are reci-
procally in a fubduplicate proportion of thickneffes of
the plates of Glafs.

So then in plates of Glafs which are alike concave on
one fide, and alike convex on the other fide, and alike
quick-filvered on the convex fides, and differ in nothing
but

but their thicknefs, the Diameters of the Rings are re-
ciprocally in a fubduplicate proportion of the thicknesses
of the plates. And this shews fufficiently that the Rings
depend on both the furfaces of the Glafs. They depend on the convex furface becaufe they are more luminous when that furface is quick-filvered over than
when it is without Quick-filver. They depend alfo
upon the concave furface, becaufe without that furface
a Speculum makes them not. They depend on both
furfaces and on the diftances between them, becaufe
their bignefs is varied by varying only that diftance.
And this dependance is of the fame kind with that
which the Colours of thin plates have on the diftance
of the furfaces of thofe plates, becaufe the bignefs
of the Rings and their proportion to one another,
and the variation of their bignefs arifing from the variation of the thicknefs of the Glafs, and the orders of
their Colours, is fuch as ought to refult from the Propofitions in the end of the third Part of this Book, derived
from the the Phænomena of the Colours of thin plates
fet down in the firft Part.

There are yet other Phænomena of thefe Rings of
Colours but fuch as follow from the fame Propofitions,
and therefore confirm both the truth of thofe Propofitions, and the Analogy between thefe Rings and the
Rings of Colours made by very thin plates. I fhall
fubjoyn fome of them.

O B S.

O B S. X.

When the beam of the Sun's Light was reflected back from the Speculum not directly to the Hole in the Window, but to a place a little distant from it, the common center of that Spot, and of all the Rings of Colours fell in the middle way between the beam of the incident Light, and the beam of the reflected Light, and by consequence in the center of the spherical concavity of the Speculum, whenever the Chart on which the Rings of Colours fell was placed at that center. And as the beam of reflected Light by inclining the Speculum receded more and more from the beam of incident Light and from the common center of the coloured Rings between them, those Rings grew bigger and bigger, and so also did the white round Spot, and new Rings of Colours emerged succeffively out of their common center, and the white Spot became a white Ring encompaffing them ; and the incident and reflected beams of Light always fell upon the oppofite parts of this Ring, illuminating its perimeter like two mock Suns in the oppofite parts of an Iris. So then the Diameter of this Ring, meafured from the middle of its Light on one fide to the middle of its Light on the other fide, was always equal to the diftance between the middle of the incident beam of Light, and the middle of the reflected beam meafured at the Chart on which the Rings appeared : And the rays which formed this Ring were reflected by the Speculum in Angles equal to their Angles of incidence, and by confequence to their Angles of refraction at their entrance into the Glafs, but yet their Angles of

P p   reflexion

reflexion were not in the fame planes with their Angles of incidence.

O B S. XI.

The Colours of the new Rings were in a contrary order to thofe of the former, and arofe after this manner. The white round Spot of Light in the middle of the Rings continued white to the center till the diftance of the incident ond reflected beams at the chart was about $\frac{7}{8}$ parts of an Inch, and then it began to grow dark in the middle. And when that diftance was about $1\frac{3}{16}$ of an Inch, the white Spot was become a Ring encompaffing a dark round Spot which in the middle inclined to violet and indico. And the luminous Rings incompaffing it were grown equal to thofe dark ones which in the four firft Obfervations encompaffed them, that is to fay, the white Spot was grown a white Ring equal to the firft of thofe dark Rings, and the firft of thofe luminous Rings was now grown equal to the fecond of thofe dark ones, and the fecond of thofe luminous ones to the third of thofe dark ones, and fo on. For the Diameters of the luminous Rings were now $1\frac{1}{16}$, $2\frac{1}{16}$, $2\frac{2}{3}$, $3\frac{1}{10}$, &c. Inches.

When the diftance between the incident and reflected beams of Light became a little bigger, there emerged out of the middle of the dark Spot after the indico a blue, and then out of that blue a pale green, and foon after a yellow and red. And when the Colour at the center was brighteft, being between yellow and red, the bright Rings were grown equal to thofe Rings which in the four firft Obfervations next encompaffed them;

that

that is to fay, the white Spot in the middle of thofe
Rings was now become a white Ring equal to the firft
of thofe bright Rings, and the firft of thofe bright ones
was now become equal to the fecond of thofe, and fo
on. For the Diameters of the white Rings, and of the
other luminous Rings incompaffing it, were now $1\frac{11}{16}$,
$2\frac{1}{8}$, $2\frac{11}{12}$, $3\frac{1}{8}$, &c. or thereabouts.

When the diftance of the two beams of Light at the
Chart was a little more increafed, there emerged out
of the middle in order after the red, a purple, a blue,
a green, a yellow, and a red inclining much to purple,
and when the Colour was brighteft being between yellow and red, the former indico, blue, green, yellow and
red, were become an Iris or Ring of Colours equal
to the firft of thofe luminous Rings which appeared in
the four firft Obfervations, and the white Ring which
was now become the fecond of the luminous Rings was
grown equal to the fecond of thofe, and the firft of
thofe which was now become the third Ring was become the third of thofe, and fo on. For their Diame-
ters were $1\frac{11}{16}$, $2\frac{1}{8}$, $2\frac{11}{12}$, $3\frac{1}{8}$ Inches, the diftance of the
two beams of Light, and the Diameter of the white
Ring being $2\frac{1}{8}$ Inches.

When thefe two beams became more diftant there
emerged out of the middle of the purplifh red, firft a
darker round Spot, and then out of the middle of that
Spot a brighter. And now the former Colours (purple,
blue, green, yellow, and purplifh red) were become a
Ring equal to the firft of the bright Rings mentioned in
the four firft Obfervations, and the Ring about this
Ring were grown equal to the Rings about that refpectively ; the diftance between the two beams of

Light

Light and the Diameter of the white Ring (which was now become the third Ring) being about 3 Inches.

The Colours of the Rings in the middle began now to grow very dilute, and if the diftance between the two beams was increafed half an Inch, or an Inch more, they vanifhed whilft the white Ring, with one or two of the Rings next it on either fide, continued ftill vifible. But if the diftance of the two beams of Light was ftill more increafed thefe alfo vanifhed : For the Light which coming from feveral parts of the Hole in the Window fell upon the Speculum in feveral Angles of incidence made Rings of feveral bigneffes, which diluted and blotted out one another, as I knew by intercepting fome part of that Light. For if I intercepted that part which was neareft to the Axis of the Speculum the Rings would be lefs, if the other part which was remoteft from it they would be bigger.

O B S. XII.

When the Colours of the Prifm were caft fucceffively on the Speculum, that Ring which in the two laft Obfervations was white, was of the fame bignefs in all the Colours, but the Rings without it were greater in the green than in the blue, and ftill greater in the yellow, and greateft in the red. And, on the contrary, the Rings within that white Circle were lefs in the green than in the blue, and ftill lefs in the yellow, and leaft in the red. For the Angles of reflexion of thofe rays which made this Ring being equal to their Angles of incidence, the fits of every reflected ray within the Glafs

after

after reflexion are equal in length and number to the fits of the fame ray within the Glafs before its incidence on the reflecting furface; and therefore fince all the rays of all forts at their entrance into the Glafs were in a fit of tranfmiffion, they were alfo in a fit of tranfmiffion at their returning to the fame furface after reflexion; and by confequence were tranfmitted and went out to the white Ring on the Chart. This is the reafon why that Ring was of the fame bignefs in all the Colours, and why in a mixture of all it appears white. But in rays which are reflected in other Angles, the intervals of the fits of the leaft refrangible being greateft, make the Rings of their Colour in their progrefs from this white Ring, either outwards or inwards, increafe or decreafe by the greateft fteps; fo that the Rings of this Colour without are greateft, and within leaft. And this is the reafon why in the laft Obfervation, when the Speculum was illuminated with white Light, the exterior Rings made by all Colours appeared red without and blue within, and the interior blue without and red within.

Thefe are the Phænomena of thick convexo-concave plates of Glafs, which are every where of the fame thicknefs. There are yet other Phænomena when thefe plates are a little thicker on one fide than on the other, and others when the plates are more or lefs concave than convex, or plano-convex, or double-convex. For in all thefe cafes the plates make Rings of Colours, but after various manners; all which, fo far as I have yet obferved, follow from the Propofitions in the end of the third part of this Book, and fo confpire to confirm the truth of thofe Propofitions. But the Phæno-

mena

mena are too various, and the Calculations whereby
they follow from thofe Propofitions too intricate to be
here profecuted. I content my felf with having profecuted this kind of Phænomena fo far as to difcover their
caufe, and by difcovering it to ratify the Propofitions
in the third Part of this Book.

O B S. XIII.

As Light reflected by a Lens quick-filvered on the
back-fide makes the Rings of Colours above defcribed, fo it ought to make the like Rings of Colours
in paffing through a drop of Water. At the firft reflexion of the rays within the drop, fome Colours ought
to be tranfmitted, as in the cafe of a Lens, and others
to be reflected back to the Eye. For inftance, if the
Diameter of a fmall drop or globule of Water be about
the 500th part of an Inch, fo that a red-making ray in
paffing through the middle of this globule has 250 fits
of eafy tranfmiffion within the globule, and that all the
red-making rays which are at a certain diftance from
this middle ray round about it have 249 fits within the
globule, and all the like rays at a certain further diftance round about it have 248 fits, and all thofe at a
certain further diftance 247 fits, and fo on ; thefe concentrick Circles of rays after their tranfmiffion, falling
on a white Paper, will make concentrick rings of red
upon the Paper, fuppofing the Light which paffes
through one fingle globule ftrong enough to be fenfible.
And, in like manner, the rays of other Colours will
make Rings of other Colours. Suppofe now that in a
fair day the Sun fhines through a thin Cloud of fuch
                                                globules

globules of Water or Hail, and that the globules are all
of the same bignefs, and the Sun feen through this Cloud
fhall appear incompaffed with the like concentrick Rings
of Colours, and the Diameter of the firft Ring of red
fhall be 7½ degrees, that of the fecond 10-degrees, that
of the third 12 degrees 33 minutes. And accordingly
as the globules of Water are bigger or lefs, the Rings
fhall be lefs or bigger. This is the Theory, and experience anfwers it. For in *June* 1692. I faw by reflexion
in a Veffel of ftagnating Water three Halos Crowns or
Rings of Colours about the Sun, like three little Rain-
bows, concentrick to his Body. The Colours of the
firft or innermoft Crown were blue next the Sun, red
without, and white in the middle between the blue
and red. Thofe of the fecond Crown were purple and
blue within, and pale red without, and green in the
middle. And thofe of the third were pale blue within, and pale red without ; thefe Crowns inclofed one
another immediately, fo that their Colours proceeded
in this continual order from the Sun outward : blue,
white, red ; purple, blue, green, pale yellow and red ;
pale blue, pale red. The Diameter of the fecond Crown
meafured from the middle of the yellow and red on one
fide of the Sun, to the middle of the fame Colour on
the other fide was 9⅓ degrees, or thereabouts. The Diameters of the firft and third I had not time to meafure,
but that of the firft feemed to be about five or fix degrees, and that of the third about twelve. The like
Crowns appear fometimes about the Moon ; for in the
beginning of the year 1664, *Febr.* 19th at night, I faw
two fuch Crowns about her. The Diameter of the firft
or innermoft was about three degrees, and that of the

fecond

second about five degrees and an half. Next about the Moon was a Circle of white, and next about that the inner Crown which was of a bluish green within next the white, and of a yellow and red without, and next about these Colours were blue and green on the inside of the outward Crown, and red on the outside of it. At the same time there appeared a Halo about 22 degrees 35′ distant from the center of the Moon. It was Elliptical, and its long Diameter was perpendicular to the Horizon verging below farthest from the Moon. I am told that the Moon has sometimes three or more concentrick Crowns of Colours incompassing one another next about her Body. The more equal the globules of Water or Ice are to one another, the more Crowns of Colours will appear, and the Colours will be the more lively. The Halo at the distance of 22½ degrees from the Moon is of another sort. By its being oval and remoter from the Moon below than above, I conclude, that it was made by refraction in some sort of Hail or Snow floating in the Air in an horizontal Posture, the refracting Angle being about 58 or 60 degrees.

THE

THE

THIRD BOOK

OF

OPTICKS.

Obſervations concerning the Inflexions of the rays of Light,
and the Colours made thereby.

GRimaldo has informed us, that if a beam of the
Sun's Light be let into a dark Room through a
very ſmall Hole, the ſhadows of things in this Light
will be larger than they ought to be if the rays went
on by the Bodies in ſtreight Lines, and that theſe ſhadows have three parallel fringes, bands or ranks of coloured Light adjacent to them.   But if the Hole be
enlarged the fringes grow broad and run into one another, ſo that they cannot be diſtinguiſhed.   Theſe broad
ſhadows and fringes have been reckoned by ſome to proceed from the ordinary refraction of the Air, but without due examination of the matter.   For the circumſtances of the Phænomenon, ſo far as I have obſerved
them, are as follows.

Q q   O B S.

O B S. I.

I made in a piece of Lead a small Hole with a Pin, whose breadth was the 42th part of an Inch. For 21 of those Pins laid together took up the breadth of half an Inch. Through this Hole I let into my darkened Chamber a beam of the Sun's Light, and found that the shadows of Hairs, Thred, Pins, Straws, and such like slender substances placed in this beam of Light, were considerably broader than they ought to be, if the rays of Light passed on by these Bodies in right Lines. And particularly a Hair of a Man's Head, whose breadth was but the 280th part of an Inch, being held in this Light, at the distance of about twelve Feet from the Hole, did cast a shadow which at the distance of four Inches from the Hair was the sixtieth part of an Inch broad, that is, above four times broader than the Hair, and at the distance of two Feet from the Hair was about the eight and twentieth part of an Inch broad, that is, ten times broader than the Hair, and at the distance of ten Feet was the eighth part of an Inch broad, that is 35 times broader.

Nor is it material whether the Hair be incompassed with Air, or with any other pellucid substance. For I wetted a polished plate of Glass, and laid the Hair in the Water upon the Glass, and then laying another polished plate of Glass upon it, so that the Water might fill up the space between the Glasses, I held them in the aforesaid beam of Light, so that the Light might pass through them perpendicularly, and the shadow of the Hair was at the same distances as big as before.

The

The fhadows of fcratches made in polifhed plates of Glafs were alfo much broader than they ought to be, and the Veins in polifhed plates of Glafs did alfo caft the like broad fhadows. And therefore the great breadth of thefe fhadows proceeds from fome other caufe than the refraction of the Air.

Let the Circle X reprefent the middle of the Hair; *Fig.* 1. A D G, B E H, C F I, three rays paffing by one fide of the Hair at feveral diftances; K N Q, L O R, M P S, three other rays paffing by the other fide of the Hair at the like diftances; D, E, F and N, O, P, the places where the rays are bent in their paffage by the Hair; G, H, I and Q, R, S, the places where the rays fall on a Paper G Q; I S the breadth of the fhadow of the Hair caft on the Paper, and T I, V S, two rays paffing to the points I and S without bending when the Hair is taken away. And it's manifeft that all the Light between thefe two rays A I and V S is bent in paffing by the Hair, and turned afide from the fhadow I S, becaufe if any part of this Light were not bent it would fall on the Paper within the fhadow, and there illuminate the Paper contrary to experience. And becaufe when the Paper is at a great diftance from the Hair, the fhadow is broad, and therefore the rays T I and V S are at a great diftance from one another, it follows that the Hair acts upon the rays of Light at a good diftance in their paffing by it. But the action is ftrongeft on the rays which pafs by at leaft diftances, and grows weaker and weaker accordingly as the rays pafs by at diftances greater and greater, as is reprefented in the Scheme: For thence it comes to pafs, that the fhadow of the Hair is much broader in proportion to the diftance of

the

the Paper from the Hair, when the Paper is nearer the Hair than when it is at a great diſtance from it.

O B S. II.

The ſhadows of all Bodies (Metals, Stones, Glaſs, Wood, Horn, Ice, &c.) in this Light were bordered with three parallel fringes or bands of coloured Light, whereof that which was contiguous to the ſhadow was broadeſt and moſt luminous, and that which was remoteſt from it was narroweſt, and ſo faint, as not eaſily to be viſible.   It was difficult to diſtinguiſh the Colours unleſs when the Light fell very obliquely upon a ſmooth Paper, or ſome other ſmooth vvhite Body, ſo as to make them appear much broader than they vvould otherwiſe do.   And then the Colours were plainly viſible in this order : The firſt or innermoſt fringe was violet and deep blue next the ſhadovv, and then light blue, green and yellovv in the middle, and red vvithout.   The ſecond fringe vvas almoſt contiguous to the firſt, and the third to the ſecond, and both vvere blue vvithin and yellovv and red vvithout, but their Colours vvere very faint eſpecially thoſe of the third.   The Colours therefore proceeded in this order from the ſhadovv, violet, indico, pale blue, green, yellovv, red ; blue, yellovv, red ; pale blue, pale yellovv and red.   The ſhadows made by ſcratches and bubbles in poliſhed plates of Glaſs vvere bordered vvith the like fringes of coloured Light. And if plates of Looking-glaſs ſloop’d off near the edges vvith a Diamond cut, be held in the ſame beam of Light, the Light which paſſes through the parallel planes of the Glaſs will be be bordered with the like fringes of Colours

lours where thofe Planes meet with the Diamond cut, and by this means there will fometimes appear four or five fringes of Colours. Let A B, C D reprefent the *Fig.* 2. parallel planes of a Looking-glafs, and B D the plane of the Diamond-cut, making at B a very obtufe Angle with the plane A B. And let all the Light between the rays E N I and F B M pafs directly through the parallel planes of the Glafs, and fall upon the Paper between I and M, and all the Light between the rays G O and H D be refracted by the oblique plane of the Diamond cut B D, and fall upon the Paper between K and L ; and the Light which paffes directly through the parallel planes of the Glafs, and falls upon the Paper between I and M, will be bordered with three or more fringes at M.

O B S. III.

When the Hair was twelve Feet diftant from the Hole, and its fhadow fell obliquely upon a flat vvhite fcale of Inches and parts of an Inch placed half a Foot beyond it, and alfo when the fhadow fell perpendicularly upon the fame fcale placed nine Feet beyond it; I meafured the breadth of the fhadow and fringes as accurately as I could, and found them in parts of an Inch as follows.

The

At the distance of	half a Foot.	nine Feet.
The breadth of the Shadow	$\frac{1}{54}$	$\frac{1}{9}$
The breadth between the middles of the brighteſt Light of the innermoſt fringes on either ſide the ſhadow	$\frac{1}{38}$ or $\frac{1}{39}$	$\frac{2}{50}$
The breadth between the middles of the brighteſt Light of the middlemoſt fringes on either ſide the ſhadow	$\frac{1}{23\frac{1}{2}}$	$\frac{4}{17}$
The breadth between the middles of the brighteſt Light of the outmoſt fringes on either ſide the ſhadow	$\frac{1}{18}$ or $\frac{1}{18\frac{1}{2}}$	$\frac{3}{10}$
The diſtance between the middles of the brighteſt Light of the firſt and ſecond fringes	$\frac{1}{120}$	$\frac{1}{21}$
The diſtance between the middles of the brighteſt Light of the ſecond and third fringes	$\frac{1}{170}$	$\frac{1}{31}$
The breadth of the luminous part (green, white, yellow and red) of the firſt fringe	$\frac{1}{170}$	$\frac{1}{32}$
The breadth of the darker ſpace between the firſt and ſecond fringes.	$\frac{1}{240}$	$\frac{1}{45}$
The breadth of the luminous part of the ſecond fringe	$\frac{1}{290}$	$\frac{1}{55}$
The breadth of the darker ſpace between the ſecond and third fringes.	$\frac{1}{340}$	$\frac{1}{63}$

Theſe

These measures I took by letting the shadow of the Hair at half a Foot distance fall so obliquely on the scale as to appear twelve times broader than vvhen it fell perpendicularly on it at the same distance, and setting down in this Table the twelfth part of the measures I then took.

O B S. IV.

When the shadovv and fringes vvere cast obliquely upon a smooth vvhite Body, and that Body was removed further and further from the Hair, the first fringe began to appear and look brighter than the rest of the Light at the distance of less than a quarter of an Inch from the Hair, and the dark line or shadovv between that and the second fringe began to appear at a less distance from the Hair than that of the third part of an Inch. The second fringe began to appear at a distance from the Hair of less than half an Inch, and the shadow between that and the third fringe at a distance less than an Inch, and the third fringe at a distance less than three Inches. At greater distances they became much more sensible, but kept very nearly the same proportion of their breadths and intervals which they had at their first appearing. For the distance between the middle of the first and middle of the second fringe, was to the distance between the middle of the second and middle of the third fringe, as three to two, or ten to seven. And the last of these two distances vvas equal to the breadth of the bright Light or luminous part of the first fringe. And this breadth vvas to the breadth of the bright Light of the second fringe as seven to four, and to the dark
interval

interval of the firſt and ſecond fringe as three to two, and to the like dark interval between the ſecond and third as two to one. For the breadths of the fringes ſeemed to be in the progreſſion of the numbers 1, $\sqrt{\frac{1}{3}}$, $\sqrt{\frac{1}{5}}$ and their intervals to be in the ſame progreſſion vvith them; that is, the fringes and their intervals together to be in the continual progreſſion of the numbers 1, $\sqrt{\frac{1}{2}}$, $\sqrt{\frac{1}{3}}$, $\sqrt{\frac{1}{4}}$, $\sqrt{\frac{1}{5}}$, or thereabouts. And theſe proportions held the ſame very nearly at all diſtances from the Hair; the dark Intervals of the fringes being as broad in proportion to the fringes at their firſt appearance as afterwards at great diſtances from the Hair, though not ſo dark and diſtinct.

OBS. V.

The Sun ſhining into my darkened Chamber through a Hole a quarter of an Inch broad; I placed at the diſtance of two or three Feet from the Hole a Sheet of Paſt-board, vvhich vvas black'd all over on both ſides, and in the middle of it had a Hole about three quarters of an Inch ſquare for the Light to paſs through. And behind the Hole I faſtened to the Paſt-board vvith Pitch the blade of a ſharp Knife, to intercept ſome part of the Light vvhich paſſed through the Hole. The planes of the Paſt-board and blade of the Knife vvere parallel to one another, and perpendicular to the rays. And vvhen they vvere ſo placed that none of the Sun's Light fell on the Paſt-board, but all of it paſſed through the Hole to the Knife, and there part of it fell upon the blade of the Knife, and part of it paſſed by its edge: I let this part of the Light vvhich paſſed by, fall on a vvhite white Paper two or three Feet beyond the Knife, and there saw two streams of faint Light shoot out both ways from the beam of Light into the shadow like the tails of Comets. But because the Sun's direct Light by its brightness upon the Paper obscured these faint streams, so that I could scarce see them, I made a little Hole in the midst of the Paper for that Light to pass through and fall on a black cloth behind it ; and then I saw the two streams plainly. They were like one another, and pretty nearly equal in length and breadth, and quantity of Light. Their Light at that end next the Sun's direct Light was pretty strong for the space of about a quarter of an Inch, or half an Inch, and in all its progress from that direct Light decreased gradually till it became insensible. The whole length of either of these streams measured upon the Paper at the distance of three Feet from the Knife was about six or eight Inches ; so that it subtended an Angle at the edge of the Knife of about 10 or 12, or at most 14 degrees. Yet sometimes I thought I saw it shoot three or four degrees further, but with a Light so very faint that I could scarce perceive it, and suspected it might (in some measure at least) arise from some other cause than the two streams did. For placing my Eye in that Light beyond the end of that stream which was behind the Knife, and looking towards the Knife, I could see a line of Light upon its edge, and that not only when my Eye was in the line of the streams, but also when it was without that line either towards the point of the Knife, or towards the handle. This line of Light appeared contiguous to the edge of the Knife, and was narrower than the Light of the innermost fringe, and

R r   narrowest

narroweft when my Eye was furtheft from the direct Light, and therefore feemed to pafs between the Light of that fringe and the edge of the Knife, and that which paffed neareft the edge to be moft bent, though not all of it.

O B S. VI.

I placed another Knife by this fo that their edges might be parallel and look towards one another, and that the beam of Light might fall upon both the Knives, and fome part of it pafs between their edges. And when the diftance of their edges was about the 400th part of an Inch the ftream parted in the middle, and left a fhadow between the two parts. This fhadow was fo black and dark that all the Light which paffed between the Knives feemed to be bent, and turned afide to the one hand or to the other. And as the Knives ftill approached one another the fhadow grew broader, and the ftreams fhorter at their inward ends which were next the fhadow, until upon the contact of the Knives the whole Light vanifhed leaving its place to the fhadow.

And hence I gather that the Light which is leaft bent, and goes to the inward ends of the ftreams, paffes by the edges of the Knives at the greateft diftance, and this diftance when the fhadow begins to appear between the ftreams is about the eight-hundredth part of an Inch. And the Light which paffes by the edges of the Knives at diftances ftill lefs and lefs is more and more bent, and goes to thofe parts of the ftreams which are further and further from the direct Light, becaufe when

when the Knives approach one another till they touch, those parts of the streams vanish last which are furthest from the direct Light.

OBS. VII.

In the fifth Observation the fringes did not appear, but by reason of the breadth of the Hole in the Window became so broad as to run into one another, and by joyning make one continued Light in the beginning of the streams. But in the sixth, as the Knives approached one another, a little before the shadow appeared between the two streams, the fringes began to appear on the inner ends of the streams on either side of the direct Light, three on one side made by the edge of one Knife, and three on the other side made by the edge of the other Knife. They were distincteft when the Knives were placed at the greatest distance from the Hole in the Window, and still became more distinct by making the Hole less, insomuch that I could sometimes see a faint lineament of a fourth fringe beyond the three above-mentioned. And as the Knives continually approached one another, the fringes grew distincter and larger until they vanished. The outmost fringe vanished first, and the middlemost next, and the innermost last. And after they were all vanished, and the line of Light which was in the middle between them was grown very broad, enlarging it self on both sides into the streams of Light described in the fifth Observation, the above-mentioned shadow began to appear in the middle of this line, and divide it along the middle into two lines of Light, and increased until the whole

Light

Light vaniſhed. This inlargement of the fringes was ſo great that the rays which go to the innermoſt fringe ſeemed to be bent above twenty times more when this fringe was ready to vaniſh, than when one of the Knives was taken away.

And from this and the former Obſervation compared, I gather, that the Light of the firſt fringe paſſed by the edge of the Knife at a diſtance greater than the eight-hundredth part of an Inch, and the Light of the ſecond fringe paſſed by the edge of the Knife at a greater diſtance than the Light of the firſt fringe did, and that of the third at a greater diſtance than that of the ſecond, and that of the ſtreams of Light deſcribed in the fifth and ſixth Obſervations paſſed by the edges of the Knives at leſs diſtances than that of any of the fringes.

O B S. VIII.

I cauſed the edges of two Knives to be ground truly ſtreight, and pricking their points into a board ſo that their edges might look towards one another, and meeting near their points contain a rectilinear Angle, I faſt-ned their handles together with Pitch to make this Angle invariable. The diſtance of the edges of the Knives from one another at the diſtance of four Inches from the angular point, where the edges of the Knives met, was the eighth part of an Inch, and therefore the Angle contained by the edges was about 1 degr. 54. The Knives thus fixed together I placed in a beam of the Sun's Light, let into my darkened Chamber through a Hole the 42th part of an Inch wide, at the diſtance

of

of ten or fifteen Feet from the Hole, and let the Light which paffed between their edges fall very obliquely upon a fmooth white Ruler at the diftance of half an Inch, or an Inch from the Knives, and there faw the fringes made by the two edges of the Knives run along the edges of the fhadows of the Knives in lines parallel to thofe edges without growing fenfibly broader, till they met in Angles equal to the Angle contained by the edges of the Knives, and where they met and joyned they ended without croffing one another. But if the Ruler was held at a much greater diftance from the Paper, the fringes became fomething broader and broader as they approached one another, and after they met they croffed one another, and then became much broader than before.

Whence I gather that the diftances at which the fringes pafs by the Knives are not increafed nor altered by the approach of the Knives, but the Angles in which the rays are there bent are much increafed by that approach ; and that the Knife which is neareft any ray determines which way the ray fhall be bent, and the other Knife increafes the bent.

O B S. IX.

When the rays fell very obliquely upon the Ruler at the diftance of the third part of an Inch from the Knives, the dark line between the firft and fecond fringe of the fhadow of one Knife, and the dark line between the firft and fecond fringe of the fhadow of the other Knife met with one another, at the diftance of the fifth part of an Inch from the end of the Light which paffed be-

tween

tween the Knives at the concourſe of their edges. And
therefore the diſtance of the edges of the Knives at the
meeting of theſe dark lines was the 160th part of an
Inch. For as four Inches to the eighth part of an Inch,
ſo is any length of the edges of the Knives meaſured
from the point of their concourſe to the diſtance of the
edges of the Knives at the end of that length, and ſo is
the fifth part of an Inch to the 160th part. So then the
dark lines above-mentioned meet in the middle of the
Light which paſſes between the Knives where they are
diſtant the 160th part of an Inch, and the one half of
that Light paſſes by the edge of one Knife at a diſtance
not greater than the 320th part of an Inch, and falling
upon the Paper makes the fringes of the ſhadow of that
Knife, and the other half paſſes by the edge of the
other Knife, at a diſtance not greater than the 320th
part of an Inch, and falling upon the Paper makes the
fringes of the ſhadow of the other Knife. But if the
Paper be held at a diſtance from the Knives greater than
the third part of an Inch, the dark lines above-mentioned meet at a greater diſtance than the fifth part of
an Inch from the end of the Light which paſſed between the Knives at the concourſe of their edges; and
therefore the Light which falls upon the Paper where
thoſe dark lines meet paſſes between the Knives
where their edges are diſtant above the 160th part of
an Inch.

For at another time when the two Knives were diſtant eight Feet and five Inches from the little Hole in
the Window, made with a ſmall Pin as above, the Light
which fell upon the Paper where the aforeſaid dark
lines met. paſſed between the Knives, where the diſtance

ftance between their edges was as in the following Table, when the diftance of the Paper from the Knives was alfo as follows.

Diftances of the Paper from the Knives in Inches.	Diftances between the edges of the Knives in mille-fimal parts of an Inch.
$1\frac{1}{2}$.	o'o12.
$3\frac{1}{3}$.	o'o20.
$8\frac{2}{5}$.	o'o34.
32.	o'o57.
96.	o'o81.
131.	o'o87.

And hence I gather that the Light which makes the fringes upon the Paper is not the fame Light at all di-ftances of the Paper from the Knives, but when the Paper is held near the Knives, the fringes are made by Light which paffes by the edges of the Knives at a lefs diftance, and is more bent than when the Paper is held at a greater diftance from the Knives.

O B S. X.

When the fringes of the fhadows of the Knives fell perpendicularly upon a Paper at a great diftance from the Knives, they were in the form of Hyperbolas, and their dimenfions were as follows. Let C A, C B repre-fent lines drawn upon the Paper parallel to the edges of the Knives, and between which all the Light would fall, if it paffed between the edges of the Knives with-out inflexion; D E a right line drawn through C making the

the Angles ACD, BCE, equal to one another, and
terminating all the Light which falls upon the Paper from
the point where the edges of the Knives meet; e i s, f k t,
and g l v, three hyperbolical lines reprefenting the terminus of the fhadow of one of the Knives, the dark line
between the firft and fecond fringes of that fhadow, and
the dark line between the fecond and third fringes of
the fame fhadow; x i p, y k q and z l r, three other Hyperbolical lines reprefenting the terminus of the fhadow
of the other Knife, the dark line between the firft and
fecond fringes of that fhadow, and the dark line between the fecond and third fringes of the fame fhadow.
And conceive that thefe three Hyperbolas are like and
equal to the former three, and crofs them in the points
i, k and l, and that the fhadows of the Knives are terminated and diftinguifhed from the firft luminous fringes
by the lines e i s and x i p, until the meeting and croffing of the fringes, and then thofe lines crofs the fringes
in the form of dark lines, terminating the firft luminous
fringes within fide, and diftinguifhing them from another Light which begins to appear at i, and illuminates
all the triangular fpace i p D E s comprehended by thefe
dark lines, and the right line DE. Of thefe Hyperbolas one Afymptote is the line DE, and their other
Afymptotes are parallel to the lines CA and CB. Let
r v reprefent a line drawn any where upon the Paper
parallel to the Afymptote DE, and let this line crofs
the right lines AC in m and BC in n, and the fix dark
hyperbolical lines in p, q, r; s, t, v; and by meafuring
the diftances p s, q t, r v, and thence collecting the
the lengths of the ordinates n p, n q, n r or m s, m t,
m v, and doing this at feveral diftances of the line r v,
from

from the Afymptote D E you may find as many points of thefe Hyperbolas as you pleafe, and thereby know that thefe curve lines are Hyperbolas differing little from the conical Hyperbola. And by meafuring the lines Ci, Ck, Cl, you may find other points of thefe Curves.

For inftance, when the Knives were diftant from the Hole in the Window ten Feet, and the Paper from the Knives 9 Feet, and the Angle contained by the edges of the Knives to which the Angle ACB is equal, was fubtended by a chord which was to the Radius as 1 to 32, and the diftance of the line r v from the Afymptote D E was half an Inch: I meafured the lines p s, q t, r v, and found them 0'35, 0'65, 0'98 Inches refpectively, and by adding to their halfs the line ‡ m n (which here was the 128th part of an Inch, or 0'0078 Inches) the fums n p, n q, n r, were 0'1828, 0'3328, 0'4978 Inches. I meafured alfo the diftances of the brighteft parts of the fringes which run between p q and s t, q r and t v, and next beyond r and v, and found them 0'5, 0'8, and 1'17 Inches.

O B S. XI.

The Sun fhining into my darkened Room through a fmall round Hole made in a plate of Lead with a flender Pin as above; I placed at the Hole a Prifm to refract the Light, and form on the oppofite Wall the Spectrum of Colours, defcribed in the third Experiment of the firft Book. And then I found that the fhadows of all Bodies held in the coloured Light between the Prifm and the Wall, were bordered with fringes of the Colour

S s   of

of that Light in which they were held.   In the full red
Light they were totally red without any fenfible blue
or violet, and in the deep blue Light they were totally
blue without any fenfible red or yellow ; and fo in the
green Light they were totally green, excepting a little
yellow and blue, which were mixed in the green Light
of the Prifm.   And comparing the fringes made in the
feveral coloured Lights, I found that thofe made in the
red Light were largeft, thofe made in the violet were
leaft, and thofe made in the green were of a middle
bignefs.   For the fringes with which the fhadow of a
Man's Hair were bordered, being meafured crofs the
fhadow at the diftance of fix Inches from the Hair ; the
diftance between the middle and moft luminous part of
the firft or innermoft fringe on one fide of the fhadow,
and that of the like fringe on the other fide of the fha-
dow, was in the full red Light $\frac{1}{37\frac{1}{2}}$ of an Inch, and in
the full violet $\frac{1}{46}$.   And the like diftance between the
middle and moft luminous parts of the fecond fringes on
either fide the fhadow was in the full red Light $\frac{1}{22\frac{1}{2}}$, and
in the violet $\frac{1}{27}$ of an Inch.   And thefe diftances of the
fringes held the fame proportion at all diftances from
the Hair without any fenfible variation.

So then the rays which made thefe fringes in the red
Light paffed by the Hair at a greater diftance than thofe
did which made the like fringes in the violet ; and therefore the Hair in caufing thefe fringes acted alike upon
the red Light or leaft refrangible rays at a greater diftance, and upon the violet or moft refrangible rays at
a lefs diftance, and by thofe actions difpofed the red
Light into larger fringes, and the violet into fmaller,
and the Lights of intermediate Colours into fringes of
inter-

intermediate bigneſſes without changing the Colour of
of any ſort of Light.

When therefore the Hair in the firſt and ſecond of
theſe Obſervations was held in the white beam of the
Sun's Light, and caſt a ſhadow which was bordered with
three fringes of coloured Light, thoſe Colours aroſe not
from any new modifications impreſt upon the rays of
Light by the Hair, but only from the various inflections
whereby the ſeveral ſorts of rays were ſeparated from
one another, which before ſeparation by the mixture
of all their Colours, compoſed the white beam of the
Sun's Light, but whenever ſeparated compoſe Lights
of the ſeveral Colours which they are originally diſpoſed to exhibit. In this 13th Obſervation, where the
Colours are ſeparated before the Light paſſes by the
Hair, the leaſt refrangible rays, which when ſeparated from the reſt make red, were inflected at a greater
diſtance from the Hair, ſo as to make three red fringes
at a greater diſtance from the middle of the ſhadow of
the Hair ; and the moſt refrangible rays which when
ſeparated make violet, were inflected at a leſs diſtance
from the Hair, ſo as to make three violet fringes at a
leſs diſtance from the middle of the ſhadow of the Hair.
And other rays of intermediate degrees of refrangibility were inflected at intermediate diſtances from the
Hair, ſo as to make fringes of intermediate Colours at
intermediate diſtances from the middle of the ſhadow
of the Hair. And in the ſecond Obſervation, where
all the Colours are mixed in the white Light which
paſſes by the Hair, theſe Colours are ſeparated by the
various inflexions of the rays, and the fringes which
they make appear all together, and the innermoſt

fringes

fringes being contiguous make one broad fringe compofed of all the Colours in due order, the violet lying
on the infide of the fringe next the fhadow, the red on
the outfide furtheft from the fhadow, and the blue,
green and yellow, in the middle. And, in like manner, the middlemoft fringes of all the Colours lying in
order, and being contiguous, make another broad fringe
compofed of all the Colours; and the outmoft fringes
of all the Colours lying in order, and being contiguous,
make a third broad fringe compofed of all the Colours.
Thefe are the three fringes of coloured Light with
which the fhadows of all Bodies are bordered in the fecond Obfervation.

When I made the foregoing Obfervations, I defigned
to repeat moft of them with more care and exactnefs,
and to make fome new ones for determining the manner how the rays of Light are bent in their paffage by
Bodies for making the fringes of Colours with the
dark lines between them. But I was then interrupted, and cannot now think of taking thefe things into
further confideration. And fince I have not finifhed
this part of my Defign, I fhall conclude, with propo-
fing only fome Queries in order to a further fearch to
be made by others.

Query 1. Do not Bodies act upon Light at a diftance,
and by their action bend its rays, and is not this action
(*cæteris paribus*) ftrongeft at the leaft diftance?

Qu. 2. Do not the rays which differ in refrangibility
differ alfo in flexibility, and are they not by their different inflexions feparated from one another, fo as
after feparation to make the Colours in the three fringes
above

above defcribed ? And after what manner are they in-flected to make thofe fringes ?

Qu. 3. Are not the rays of Light in paffing by the edges and fides of Bodies, bent feveral times backwards and forwards, with a motion like that of an Eel ? And do not the three fringes of coloured Light above-mentioned, arife from three fuch bendings?

Qu. 4. Do not the rays of Light which fall upon Bodies, and are reflected or refracted, begin to bend before they arrive at the Bodies ; and are they not reflected, refracted and inflected by one and the fame Principle, acting varioufly in various circumftances?

Qu. 5. Do not Bodies and Light act mutually upon one another, that is to fay, Bodies upon Light in emitting, reflecting, refracting and inflecting it, and Light upon Bodies for heating them, and putting their parts into a vibrating motion wherein heat confifts ?

Qu. 6. Do not black Bodies conceive heat more eafily from Light than thofe of other Colours do, by reafon that the Light falling on them is not reflected outwards, but enters the Bodies, and is often reflected and refracted within them, until it be ftifled and loft ?

Qu. 7. Is not the ftrength and vigor of the action between Light and fulphureous Bodies obferved above, one reafon why fulphureous Bodies take fire more readily, and burn more vehemently , then other Bodies do ?

Qu. 8. Do not all fixt Bodies when heated beyond a certain degree, emit Light and fhine, and is not this emiffion performed by the vibrating motions of their parts ?

Qu. 9.

Qu. 9. Is not fire a Body heated fo hot as to emit Light copioufly? For what elfe is a red hot Iron than fire? And what elfe is a burning Coal than red hot Wood?

Qu. 10. Is not flame a vapour, fume or exhalation heated red hot, that is, fo hot as to fhine? For Bodies do not flame without emitting a copious fume, and this fume burns in the flame. The *Ignis Fatuus* is a vapour fhining without heat, and is there not the fame difference between this vapour and flame, as between rotten Wood fhining without heat and burning Coals of fire? In diftilling hot Spirits, if the head of the ftill be taken off, the vapour which afcends out of the Still will take fire at the flame of a Candle, and turn into flame, and the flame will run along the vapour from the Candle to the Still. Some Bodies heated by motion or fermentation, if the heat grow intenfe fume copioufly, and if the heat be great enough the fumes will fhine and become flame. Metals in fufion do not flame for want of a copious fume, except Spelter which fumes copioufly, and thereby flames. All flaming Bodies, as Oyl, Tallow, Wax, Wood, foffil Coals, Pitch, Sulphur, by flaming wafte and vanifh into burning fmoke, which fmoke, if the flame be put out, is very thick and vifible, and fometimes fmells ftrongly, but in the flame lofes its fmell by burning, and according to the nature of the fmoke the flame is of feveral Colours, as that of Sulphur blue, that of Copper opened with Sublimate green, that of Tallow yellow. Smoke paffing through flame cannot but grow red hot, and red hot fmoke can have no other appearance than that of flame.

Qu. 11.

Qu. 11. Do not great Bodies conferve their heat the longeft, their parts heating one another, and may not great denfe and fix'd Bodies, when heated beyond a certain degree, emit Light fo copioufly, as by the emiffion and reaction of its Light, and the reflexions and refractions of its rays within its pores to grow ftill hotter, till it comes to a certain period of heat, fuch as is that of the Sun? And are not the Sun and fix'd Stars great Earths vehemently hot, whofe heat is conferved by the greatnefs of the Bodies, and the mutual action and reaction between them, and the Light which they emit, and whofe parts are kept from fuming away, not only by their fixity, but alfo by the vaft weight and denfity of the Atmofpheres incumbent upon them, and very ftrongly compreffing them, and condenfing the vapours and exhalations which arife from them?

Qu. 12. Do not the rays of Light in falling upon the bottom of the Eye excite vibrations in the *Tunica retina?* Which vibrations, being propagated along the folid fibres of the optick Nerves into the Brain, caufe the fenfe of feeing. For becaufe denfe Bodies conferve their heat a long time, and the denfeft Bodies conferve their heat the longeft, the vibrations of their parts are of a lafting nature, and therefore may be propagated along folid fibres of uniform denfe matter to a great diftance, for conveying into the Brain the impreffions made upon all the Organs of fenfe. For that motion which can continue long in one and the fame part of a Body, can be propagated a long way from one part to another, fuppofing the Body homogeneal, fo that the motion may not be reflected, refracted, interrupted or difordered by any unevennefs of the Body.

Qu. 13.

Qu. 13. Do not feveral fort of rays make vibrations of feveral bigneffes, which according to their bigneffes excite fenfations of feveral Colours, much after the manner that the vibrations of the Air, according to their feveral bigneffes excite fenfations of feveral founds? And particularly do not the moſt refrangible rays excite the ſhorteſt vibrations for making a fenfation of deep violet, the leaſt refrangible the largeſt for making a fenfation of deep red, and the feveral intermediate forts of rays, vibrations of feveral intermediate bigneſſes to make fenfations of the feveral intermediate Colours?

Qu. 14. May not the harmony and difcord of Colours arife from the proportions of the vibrations propagated through the fibres of the optick Nerves into the Brain, as the harmony and difcord of founds arifes from the proportions of the vibrations of the Air? For fome Colours are agreeable, as thofe of Gold and Indico, and others difagree.

Qu. 15. Are not the Species of Objects feen with both Eyes united where the optick Nerves meet before they come into the Brain, the fibres on the right fide of both Nerves uniting there, and after union going thence into the Brain in the Nerve which is on the right fide of the Head, and the fibres on the left fide of both Nerves uniting in the fame place, and after union going into the Brain in the Nerve which is on the left fide of the Head, and thefe two Nerves meeting in the Brain in fuch a manner that their fibres make but one entire Species or Picture, half of which on the right fide of the Senforium comes from the right fide of both Eyes through the right fide of
both

both optick Nerves to the place where the Nerves meet, and from thence on the right fide of the Head into the Brain, and the other half on the left fide of the Senforium comes in like manner from the left fide of both Eyes. For the optick Nerves of fuch Animals as look the fame way with both Eyes (as of Men, Dogs, Sheep, Oxen, &c.) meet before they come into the Brain, but the optick Nerves of fuch Animals as do not look the fame way with both Eyes (as of Fifhes and of the Chameleon) do not meet, if I am rightly informed.

Qu. 16. When a Man in the dark preffes either corner of his Eye with his Finger, and turns his Eye away from his Finger, he will fee a Circle of Colours like thofe in the Feather of a Peacock's Tail? Do not thefe Colours arife from fuch motions excited in the bottom of the Eye by the preffure of the Finger, as at other times are excited there by Light for caufing Vifion? And when a Man by a ftroke upon his Eye fees a Flafh of Light, are not the like Motions excited in the *Retina* by the ftroke?

T t

Fig. 1.

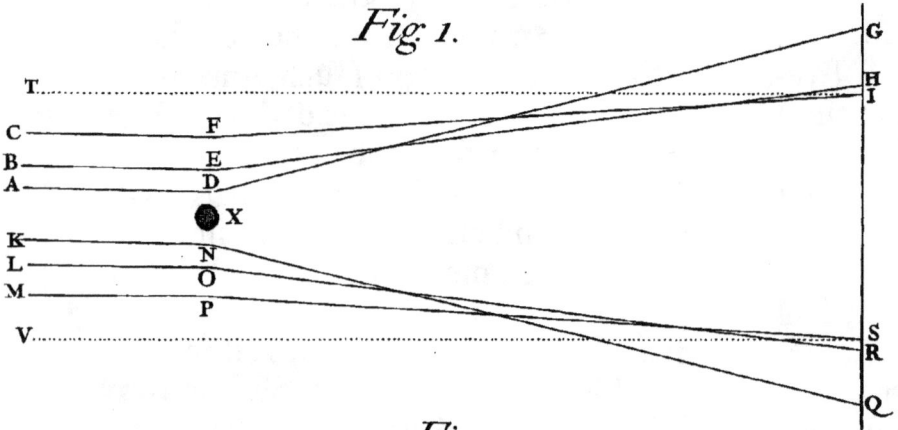

Fig. 2.

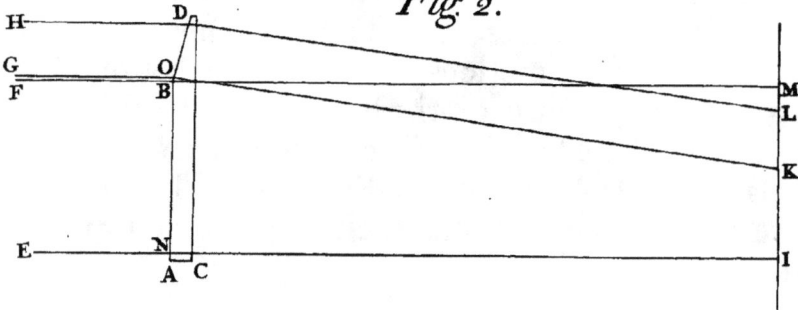

Fig. 3.

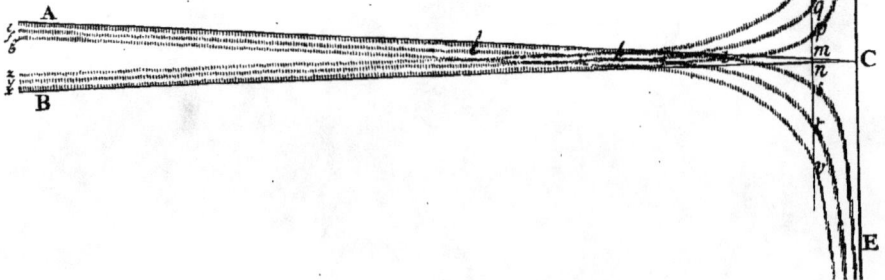

ENUMERATIO

LINEARUM

TERTII ORDINIS.

ENUMERATIO
LINEARUM
TERTII ORDINIS.

Lineæ Geometricæ secundum numerum dimen-
sionum æquationis qua relatio inter Ordinatas
& Absciffas definitur, vel (quod perinde eft) secundum numerum punctorum in quibus a linea recta
secari poffunt, optimè diftinguuntur in Ordines.
Qua ratione linea primi Ordinis erit Recta fola, eæ
secundi five quadratici ordinis erunt sectiones Conicæ
& Circulus, & eæ tertii five cubici Ordinis Parabola
Cubica, Parabola Neiliana, Ciffois veterum & reliquæ quas hic enumerare fuscepimus.   Curva autem
primi generis, (fiquidem recta inter Curvas non eft
numeranda) eadem eft cum Linea secundi Ordinis,
& Curvà secundi generis eadem cum Linea Ordinis
tertii.   Et Linea Ordinis infinitesimi ea eft quam
recta in punctis infinitis secare poteft, qualis eft Spi-
ralis, Cyclois, Quadratrix & linea omnis quæ per
radii vel rotæ revolutiones infinitas generatur.

1.
Linearum Or-
dines.

Tt 2   Sectionum

II.
*Proprietates Se-
ctionem Conica-
rum competunt
curvis superiorum
generum.*
Sectionum Conicarum proprietates præcipuæ a Geometris paffim traduntur. Et confimiles funt proprietates Curvarum fecundi generis & reliquarum, ut ex fequenti proprietatum præcipuarum enumeratione conftabit.

III.
*Curvarum fe-
cundi generis Or-
dinata, Diame-
tri, Vertices, Cen-
tra, Axes.*
Nam fi rectæ plures parallelæ & ad conicam fectionem utrinq; terminatæ ducantur, recta duas earum bifecans bifecabit alias omnes, ideoq; dicitur *Diameter* figuræ & rectæ bifectæ dicuntur *Ordinatim applicatæ* ad Diametrum, & concurfus omnium Diametrorum eft *Centrum* figuræ, & interfectio Curvæ & diametri *Vertex* nominatur, & diameter illa *Axis* eft cui ordinatim applicatæ infiftunt ad angulos rectos. Et ad eundem modum in Curvis fecundi generis, fi rectæ duæ quævis parallelæ ducantur occurrentes Curvæ in tribus punctis : recta quæ ita fecat has parallelas ut fumma duarum partium ex uno fecantis latere ad curvam terminatarum æquetur parti tertiæ ex altero latere ad curvam terminatæ, eodem modo fecabit omnes alias his parallelas curvæq; in tribus punctis occurrentes rectas, hoc eft, ita ut fumma partium duarum ex uno ipfius latere femper æquetur parti tertiæ ex altero latere. Has itaq; tres partes quæ hinc inde æquantur, *Ordinatim applicatas* & rectam fecantem cui ordinatim applicantur *Diametrum* & interfectionem diametri & curvæ *Verticem* & concurfum duarum diametrorum *Centrum* nominare licet. Diameter autem ad Ordinatas rectangula fi modo aliqua fit, etiam *Axis* dici poteft, & ubi omnes diametri in eodem puncto concurrunt iftud erit *Centrum generale.*

Hyper-

Hyperbola primi generis duas *Afymptoios*, ea fecundi tres, ea tertii quatuor & non plures habere poteft, & fic in reliquis. Et quemadmodum partes
lineæ cujufvis rectæ inter Hyperbolam Conicam &
duas ejus Afymptotos funt hinc inde æquales : fic in
Hyperbolis fecundi generis fi ducatur recta quævis
fecans tam Curvam quàm tres ejus Afymptotos in
tribus punctis, fumma duarum partium iftius rectæ
quæ a duobus quibufvis Afymptotis in eandem plagam ad duo puncta Curvæ extenduntur æqualis erit
parti tertiæ quæ a tertia Afymptoto in plagam contrariam ad tertium Curvæ punctum extenditur.

Et quemadmodum in Conicis fectionibus non Pa-
rabolicis quadratum Ordinatim applicatæ, hoc eft,
rectangulum Ordinatarum quæ ad contrarias partes Diametri ducuntur, eft ad rectangulum partium
Diametri quæ ad Vertices Ellipfeos vel Hyperbolæ
terminantur, ut data quædam linea quæ dicitur *Latus
rectum*, ad partem diametri quæ inter Vertices jacet
& dicitur *Latus tranfverfum* : fic in Curvis non Para-
bolicis fecundi generis Parallelepipedum fub tribus
Ordinatim applicatis eft ad Parallelepipedum fub partibus Diametri ad Ordinatas & tres Vertices figuræ ab-
fciffis, in ratione quadam data : in qua ratione fi fumantur tres rectæ ad tres partes diametri inter vertices figuræ fitas fingulæ ad fingulas, tunc illæ tres
rectæ dici poffunt *Latera recta* figuræ, & illæ partes
Diametri inter Vertices *Latera tranfverfa*. Et ficut
in Parabola Conica quæ ad unam & eandem diametrum unicum tantum habet Verticem, rectangulum
fub Ordinatis æquatur rectangulo fub parte Diametri
quæ ad Ordinatas & Verticem abfcinditur & recta
quadam

quadam data quæ Latus rectum dicitur,sic in Curvis
secundi generis quæ non nisi duos habent Vertices ad
eandem Diametrum, Parallelepipedum sub Ordinatis
tribus æquatur Parallelepipedo sub duabus partibus
Diametri ad Ordinatas & Vertices illos duos absciffis,
& recta quadam data quæ proinde *Latus rectum*
dici potest.

VI.
Ratio contento-
rum sub Paralle-
larum segmentis.

Deniq; ficut in Conicis sectionibus ubi duæ parallelæ ad Curvam utrinq; terminatæ secantur a duabus parallelis ad Curvam utrinq; terminatis, prima
a tertia & secunda a quarta, rectangulum partium
primæ est ad rectangulum partium tertiæ ut rectangulum partium secundæ ad rectangulum partium
quartæ: fic ubi quatuor tales rectæ occurrunt Curvæ
secundi generis fingulæ in tribus punctis, parallelepipedum partium primæ rectæ erit ad parallelepide-
dum partium tertiæ, ut parallelepipedum partium
secundæ ad parallelepipedum partium quartæ.

VII.
Crura Hyper-
bolica & Parabo-
lica & eorum pla-
gæ.

Curvarum secundi & superiorum generum æque
atq; primi crura omnia in infinitum progredientia
vel *Hyperbolici* funt generis vel *Parabolici*. Crus *Hy-*
perbolicum voco quod ad Asymptoton aliquam in infinitum appropinquat, *Parabolicum* quod Asymptoto
destituitur. Hæc crura ex tangentibus optime dig-
noscuntur. Nam fi punctum contactus in infinitum
abeat tangens cruris Hyperbolici cum Asymptoto
coincidet & tangens cruris Parabolici in infinitum
recedet, evanescet & nullibi reperietur. Invenitur
igitur Asymptotos cruris cujusvis quærendo tangen-
tem cruris illius ad punctum infinite diftans. Plaga
autem cruris infiniti invenitur quærendo positionem
rectæ cujusvis quæ tangenti parallela est ubi punctum

&ct;um conta&ct;us in infinitum abit. Nam hæc re&ct;a in eandem plagam cum crure infinito dirigitur.

Lineæ omnes Ordinis primi, tertii, quinti, fep-
timi & imparis cujufq; duo habent ad minimum
crura in infinitum verfus plagas oppofitas progredientia. Et lineæ omnes tertii Ordinis duo habent
ejufmodi crura in plagas oppofitas progredientia in
quas nulla alia earum crura infinita (præterquam
in Parabola Cartefiana) tendunt. Si crura illa
fint Hyperbolici generis, fit G A S eorum Afymptotos & huic parallela agatur re&ct;a quævis C B c
ad Curvam utrinque (fi fieri poteft) terminata
eademq; bifecetur in pun&ct;o X, & locus pun&ct;i il- *Fig.* 1,
lius X erit Hyperbola Conica (puta X Φ) cujus
una Afymptotos eft A S. Sit ejus altera Afymptotos A B, & æquatio qua relatio inter Ordinatam
B C & Abfciffam A B definitur, fi A B dicatur x &
B C y, femper induet hanc formam $xyy + ey = ax^3$
$+ bxx + cx + d$. Ubi termini e, a, b, c, d, defignant quantitates datas cum fignis fuis + & — affe-
&ct;as, quarum quælibet deeffe poffunt modo ex earum
defe&ct;u figura in fe&ct;ionem conicam non vertatur.
Poteft autem Hyberbola illa Conica cum afymptotis fuis coincidere, id eft pun&ct;um X in re&ct;a A B
locari: & tunc terminus + e y deeft.

At fi re&ct;a illa C B c non poteft utrinq; ad Curvam
terminari fed Curvæ in unico tantum pun&ct;o occur-
rit: age quamvis pofitione datam re&ct;am A B afymptoto A S occurrentem in A, ut & aliam quamvis B C
afymptoto illi parallelam Curvæque occurrentem in
pun&ct;o C, & æquatio qua relatio inter Ordinatam
BC

B C & Abſciſſam A B definitur, ſemper induet hanc formam $x y = a x^3 + b x x + c x + d$.

X.
Caſus tertius.

Quod ſi crura illa oppoſita Parabolici ſint generis, recta C B c ad Curvam utrinque, ſi fieri poteſt, terminata in plagam crurum ducatur & biſecetur in B, & locus puncti B erit linea recta. Sit iſta A B, terminata ad datum quodvis punctum A, & æquatio qua relatio inter Ordinatam BC & Abſciſſam A B definitur, ſemper induet hanc formam, $y y = a x^3 + b x x + c x + d$.

XI.
Caſus quartus.

At vero ſi recta illa C B c in unico tantum puncto occurrat Curvæ, ideoq; ad Curvam utrinq; terminari non poſſit: ſit punctum illud C, & incidat recta illa ad punctum B in rectam quamvis aliam poſitione datam & ad datum quodvis punctum A terminatam A B: & æquatio qua relatio inter Ordinatam B C & Abſciſſam AC definitur ſemper induet hanc formam, $y = a x^3 + b x x + c x + d$.

XII.
Nomina formarum.

Enumerando curvas horum caſuum, Hyperbolam vocabimus *inſcriptam* quæ tota jacet in Aſymptotôn angulo ad inſtar Hyperbolæ conicæ, *circumſcriptam* quæ Aſymptotos ſecat & partes abſciſſas in ſinu ſuo amplectitur, *ambigenam* quæ uno crure infinito inſcribitur & altero circumſcribitur, *convergentem* cujus crura concavitate ſua ſeinvicem reſpiciunt & in plagam eandem diriguntur, *divergentem* cujus crura convexitate ſua ſeinvicem recipiunt & in plagas contrarias diriguntur, *cruribus contrariis præditam* cujus crura in partes contrarias convexa ſunt & in plagas contrarias infinita, *Conchoidalem* quæ vertice concavo & cruribus divergentibus ad aſymptoton applicatur, *anguineam* quæ flexibus contrariis aſymptoton ſecat

&

& utrinq; in crura contraria producitur, *cruciformem*
quæ conjugatam decuſſat, *nodatam* quæ ſeipſam decuſſat in orbem redeundo, *cuſpidatam* cujus partes
duæ in angulo contaɕtus concurrunt & ibi terminantur, *punɕtatam* quæ conjugatam habet Ovalem infinite parvam id eſt punɕtum, & *puram* quæ per im-
poſſibilitatem duarum radicum Ovali, Nodo, Cuſ-
pide & Punɕto conjugato privatur. Eodem ſenſu
Parabolam quoq; *convergentem, divergentem, cruri-*
bus contrariis præditam, cruciformem, nodatam, cuſ-
pidatam, punɕtatam & *puram* nominabimus.

 In caſu primo ſi. terminus a x³ affirmativus eſt Fi- XIII.
De Hyberbola
redundante &
ejus tribus A-
ſymptotis.
gura erit Hyperbola triplex cum ſex cruribus Hy-
perbolicis quæ juxta tres Aſymptotos quarum nullæ
ſunt parallelæ in infinitum progrediuntur, binæ juxta
unamquamq; in plagas contrarias. Et hæ Aſymptoti ſi terminus b x x non deeſt ſe mutuo ſecabunt
in tribus punɕtis triangulum (D d ♂) inter ſe con-
tinentes, ſin terminus b x x deeſt convergent omnes
ad idem punɕtum. In priori caſu cape A D =
$\frac{-b}{2a}$, & A d = A ♂ = $\frac{b}{2\sqrt{d}}$, ac junge D d, D ♂, & erunt
A D, D d, D ♂ tres Aſymptoti. In poſteriori duc
ordinatam quamvis B C, & in ea utrinq; produɕta
cape hine inde B F & B f ſibi mutuo æquales &
in ea ratione ad A B quam habet $\sqrt{}$d ad a, jungeq;
A F, A f, & erunt A B, A F, A f tres Aſympoti.
Hanc autem Hyperbolam vocamus redundantem
quia numero crurum Hyperbolicorum Seɕtiones Conicas ſuperat.

 In Hyperbola omni redundante ſi neq; terminus XIV.
De hujus Hy-
perbolæ diametris
& ſitu crurum
infinitorum.
e y deſit neq; ſit b b − 4 a c æquale ± a e $\sqrt{}$a curva nul-
lam habebit diametrum, ſin eorum alterutrum ac-
U u cidat

cidat curva habebit unicam diametrum, & tres fi utrumque. Diameter autem femper tranfit per interfectionem duarum Afymptoton & bifecat rectas omnes quæ ad Afymptotos illas utrinq; terminantur & parallelæ funt & Afymptoto tertiæ. Eftq; abfciffa AB diameter Figuræ quoties terminus e y deeft. Diametrum vero abfolute dictam hic & in fequentibus in vulgari fignificatu ufurpo, nempe pro abfciffa quæ paffim habet ordinatas binas æquales ad idem punctum hinc inde infiftentes.

XV.
Hyperbolæ novem redundantes quæ diametro deftituuntur & tres habent Afymptotos triangulum fapientes.

Si Hyperbola redundans nullam habet diametrum quærantur Æquationis hujus $a x^4 + b x^3 + c x x + d x + \frac{:}{:} = 0$ radices quatuor feu valores ipfius x. Eæ funto A P, A ϖ, A π, A p. Erigantur ordinatæ P T, $\varpi\tau$, $\pi\tau$, p t, & hæ tangent Curvam in punctis totidem T, τ, τ, t, & tangendo dabunt limites Curvæ per quos fpecies ejus innotefcet.

Fig. 1, 2.

Nam fi radices omnes A P, A ϖ, A π, A p funt reales, ejufdem figni & inæquales, Curva conftat ex tribus Hyperbolis, (infcripta circumfcripta & ambigena) *cum Ovali*. Hyperbolarum una jacet verfus D, altera verfus d, tertia verfus ∂, & Ovalis femper jacet intra triangulam D d ∂, atq; etiam inter medios limites τ & τ, in quibus utiq; tangitur ab ordinatis $\pi\tau$ & $\varpi\tau$. Et hæc eft fpecies prima.

Fig. 3, 4.

Si e radicibus duæ maximæ A π, A *p*, vel duæ minimæ A P, A ϖ æquantur inter fe, & ejufdem funt figni cum alteris duobus, Ovalis & Hyperbola circumfcripta fibi inxicem junguntur coeuntibus earum punctis contactus τ & t vel T & τ, & crura Hyperbolæ fefe decuffando in Ovalem continuantur, figuram *nodatam* efficientia. Quæ fpecies eft fecunda.

Si

Si e radicibus tres maximæ A*p*, A *π*, A *ϖ*, vel tres *Fig. 5, 6.*
minimæ A *π*, A *ϖ*, A P æquentur inter se, Nodus in
cuspidem acutissimum convertetur.   Nam crura duo
Hyperbolæ circumscriptæ ibi in angulo contactus
concurrent & non ultra producentur.   Et hæc est
species tertia.

Si e radicibus duæ mediæ A *ϖ* & A *π* æquentur in- *Fig. 7.*
ter se, puncta contactus *τ* & *ϒ* coincidunt, & propterea Ovalis interjecta in punctum evanuit, & constat
figura ex tribus Hyperbolis, inscripta, circumscripta
& ambigena cum *puncto* conjugato.   Quæ est species
quarta.

Si duæ ex radicibus sunt impossibiles & reliquæ *Fig. 7,8,13,14.*
duæ inæquales & ejusdem signi (nam signa contraria
habere nequeunt,) *puræ* habebuntur Hyperbolæ tres
sine Ovali vel Nodo vel cuspide vel puncto conjugato, & hæ Hyperbolæ vel ad latera trianguli ab
Asymptotis comprehensi vel ad angulos ejus jacebunt
& perinde speciem vel quintam vel sextam consti-
tuent.

Si e radicibus duæ sunt æquales & alteræ duæ *Fig. 9,10,15,16.*
vel impossibiles sunt vel reales cum signis quæ a signis æqualium radicum diversa sunt, figura *crucifor-*
mis habebitur, nempe duæ ex Hyperbolis seinvicem
decussabunt idq; vel ad verticem trianguli ab A-
symptotis comprehensi, vel ad ejus basem.   Quæ
duæ species sunt septima & octava.

Si deniq; radices omnes sunt impossibiles vel si *Fig. 11, 12.*
omnes sunt reales & inæquales & earum duæ sunt
affirmativæ & alteræ duæ negativæ, tunc duæ habebuntur Hyperbolæ ad angulos oppositos duarum

Asymp-

Afymptotòn cum Hyperbola *anguinea* circa Afymptoton tertiam. Quæ fpecies eft nona.

Et hi funt omnes radicum cafus poffibiles. Nam fi duæ radices funt æquales inter fe, & aliæ duæ funt etiam inter fe æquales, Figura evadet Sectio Conica cum linea recta.

XVI.
Hyperbola duodecim redundantes cum unica tantum Diametro.

Si Hyperbola redundans habet unicam tantum Diametrum fit ejus Diameter Abfciffa A B, & æquationis hujus $ax^3 + bxx + cx + d = o$ quære tres radices feu valores x.

Fig. 17.

Si radices illæ funt omnes reales & ejufdem figni, Figura conftabit ex *Ovali* intra triangulum D d ♂ jacente & tribus Hyperbolis ad angulos ejus, nempe circumfcripta ad angulum D & infcriptis duabus ad angulos d & ♂. Et hæc eft fpecies decima.

Fig. 18.

Si radices duæ majores funt æquales & tertia ejufdem figni, crura Hyperbolæ jacentis verfus D fefe decuffabunt in forma *Nodi* propter contactum Ovalis. Quæ fpecies eft undecima.

Fig. 19.

Si tres radices funt æquales, Hyperbola ifta fit *cufpidata* fine Ovali. Quæ fpecies eft duodecima.

Fig. 20.

Si radices duæ minores funt æquales & tertia ejufdem figni, Ovalis in *punctum* evanuit. Quæ fpecies eft decima tertia. In fpeciebus quatuor noviffimis Hyperbola quæ jacet verfus D Afymptotos in finu fuo amplectitur, reliquæ duæ in finu Afymptotòn jacent.

Fig. 20.
Fig. 21.
Fig. 22.
Fig. 23.

Si duæ ex radicibus funt impoffibiles habebuntur tres Hyperbolæ *puræ* fine Ovali decuffatione vel cufpide. Et hujus cafus fpecies funt quatuor, nempe decima quarta fi Hyperbola circumfcripta jacet verfus D &

decima

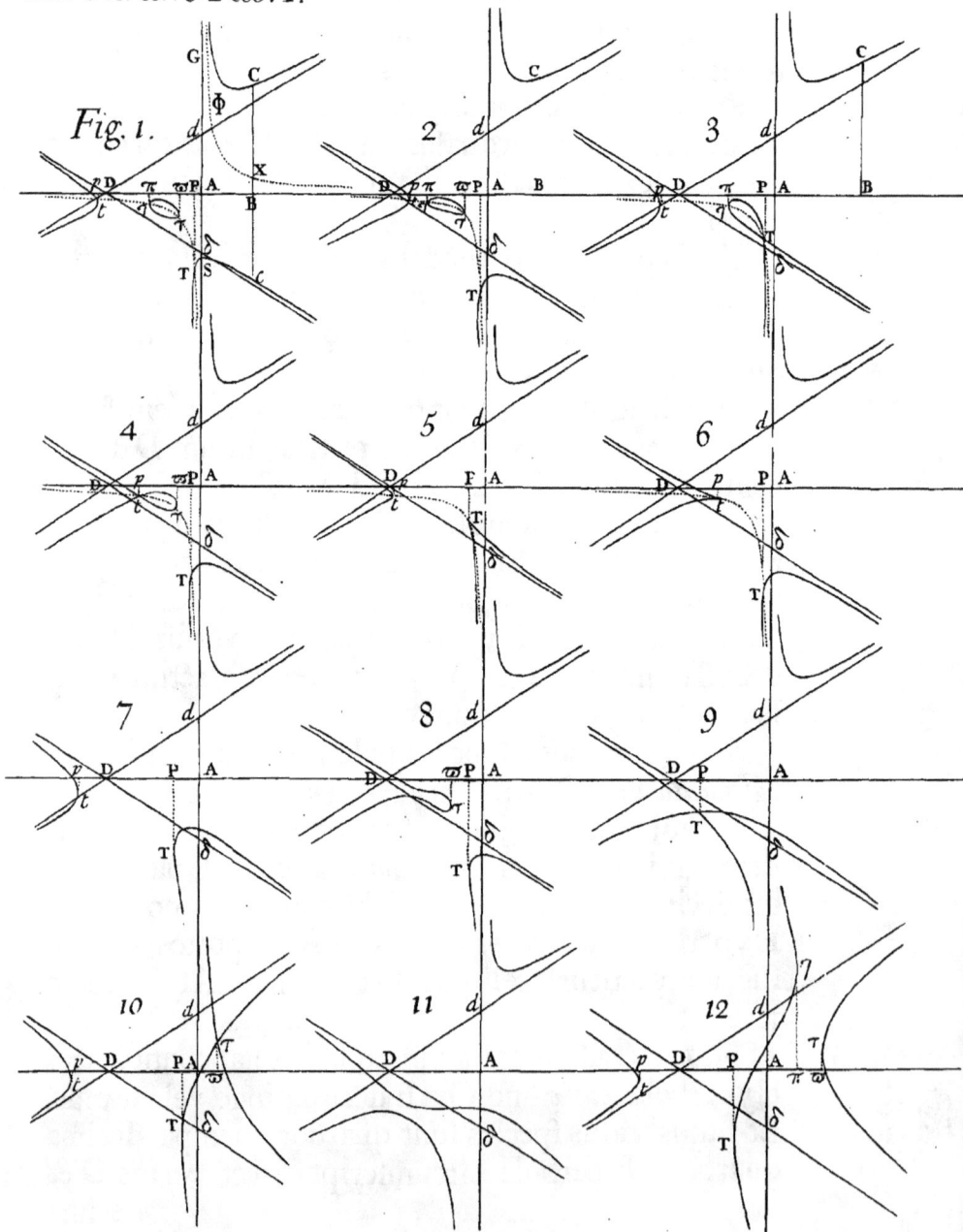

Curvarum Tab. I.

Fig. 1.   2   3   4   5   6   7   8   9   10   11   12

decima quinta fi Hyperbola infcripta jacet verfus D,
decima fexta fi Hyperbola circumfcripta jacet fub
bafi dᵈ trianguli Ddᵟ, & decima feptima fi Hyperbola infcripta jacet fub eadem bafi.

Si duæ radices funt æquales & tertia figni diverfi *Fig.* 24.
figura erit *cruciformis.* Nempe duæ ex tribus Hy- *Fig.* 25.
perbolis feinvicem decuffabunt idq; vel ad verticem
trianguli ab Afymptotis comprehenfi vel ad ejus ba-
fem. Quæ duæ fpecies funt decima octava & decima
nona.

Si duæ radices funt inæquales & ejufdem figni &
tertia eft figni diverfi, duæ habebuntur Hyperbolæ
in oppofitis angulis duarum afymptotôn cum *Con-*
choidali intermedia. Conchoidalis autem vel jace- *Fig.* 27.
bit ad eafdem partes afymptoti fuæ cum triangulo *Fig.* 26.
ab afymptotis conftituto, vel ad partes contrarias ;
& hi duo cafus conftituunt fpeciem vigefimam & vigefimam primam.

Hyperbola redundans quæ habet tres diametros   XVII.
conftat ex tribus Hyperbolis in finubus afymptotôn *Hyperbolæ duæ*
jacentibus, idq; vel ad angulos trianguli ab afympto- *redundantes cum*
tis comprehenfi vel ad ejus latera. Cafus prior dat *tribus Diametris.*
fpeciem vigefimam fecundam,& pofterior fpeciem vi- *Fig.* 28.
gefimam tertiam. *Fig.* 29.

Si tres afymptoti in puncto communi fe mutuo   XVIII.
decuffant, vertuntur fpecies quinta & fexta in vige- *Hyperbolæ no-*
fimam quartam, feptima & octava in vigefimam *vem redundantes*
quintam, & nona in vigefimam fextam ubi Anguinea *cum Afymptotis*
non tranfit per concurfum afymptotôn, & in vigefi- *tribus ad commu-*
mam feptimam ubi tranfit per concurfum illum, quo *ne punctum con-*
cafu termini b ac d defunt, & concurfus afympto- *vergentibus.*
ton eft centrum figuræ ab omnibus ejus partibus *Fig.* 30. *Fig.* 31. *Fig.* 32. *Fig.* 33.
                              oppofitis

oppofitis æqualiter diftans. Et hæ quatuor fpecies Diametrum non habent.

Fig. 34.
Fig. 35.
Fig. 36.
Fig. 37.

Vertuntur etiam fpecies decima quarta ac decima fexta in vigefimam octavam, decima quinta ac decima feptima in vigefimam nonam, decima octava & decima nona in tricefimam, & vigefima cum vigefima prima in tricefimam primam. Et hæ fpecies unicam habent diametrum.

Fig. 38.

Ac deniq; fpecies vigefima fecunda & vigefima tertia vertuntur in fpeciem tricefimam fecundam cujus tres funt Diametri per concurfum afymptotôn tranfeuntes. Quæ omnes converfiones facillime intelliguntur faciendo ut triangulum ab afymptotis comprehenfum diminuatur donec in punctum evanefcat.

XIX.
Hyperbolæ fex defectivæ diametrum non habentes.

Si in primo æquationum cafu terminus ax^3 negativus eft, Figura erit Hyberbola defectiva unicam habens afymptoton & duo tantum crura Hyperbolica juxta afymptoton illam in plagas contrarias infinite progredientia. Et afymptotos illa eft Ordinata prima & principalis AG. Si terminus e y non deeft figura nullam habebit Diametrum, fi deeft habebit unicam. In priori cafu fpecies fic enumerantur.

Fig. 39.

Si æquationis hujus ax^4=bx^3+cxx+dx+$\frac{1}{4}$ee, radices omnes Aπ, A P, Ap, Aϖ, funt reales & inæquales, Figura erit Hyperbola anguinea afymptoton flexu contrario amplexa, cum *Ovali* conjugata. Quæ fpecies eft tricefima tertia.

Fig. 40.

Si radices duæ mediæ A P & A p æquentur inter fe, Ovalis & Anguinea junguntur fefe decuffantes in forma *Nodi.* Quæ eft ipecies tricefima quarta.

Si

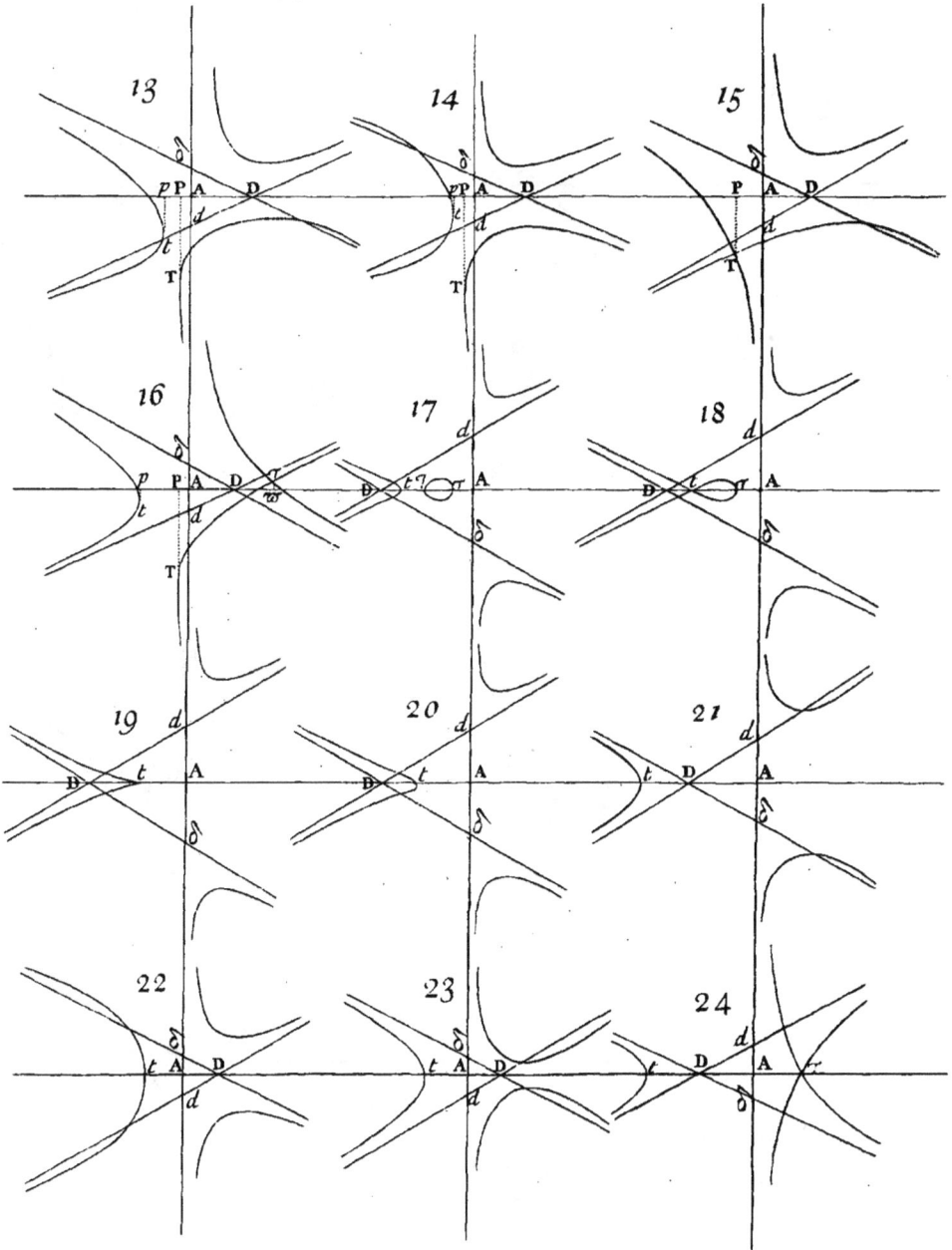

Curvarum Tab. II.

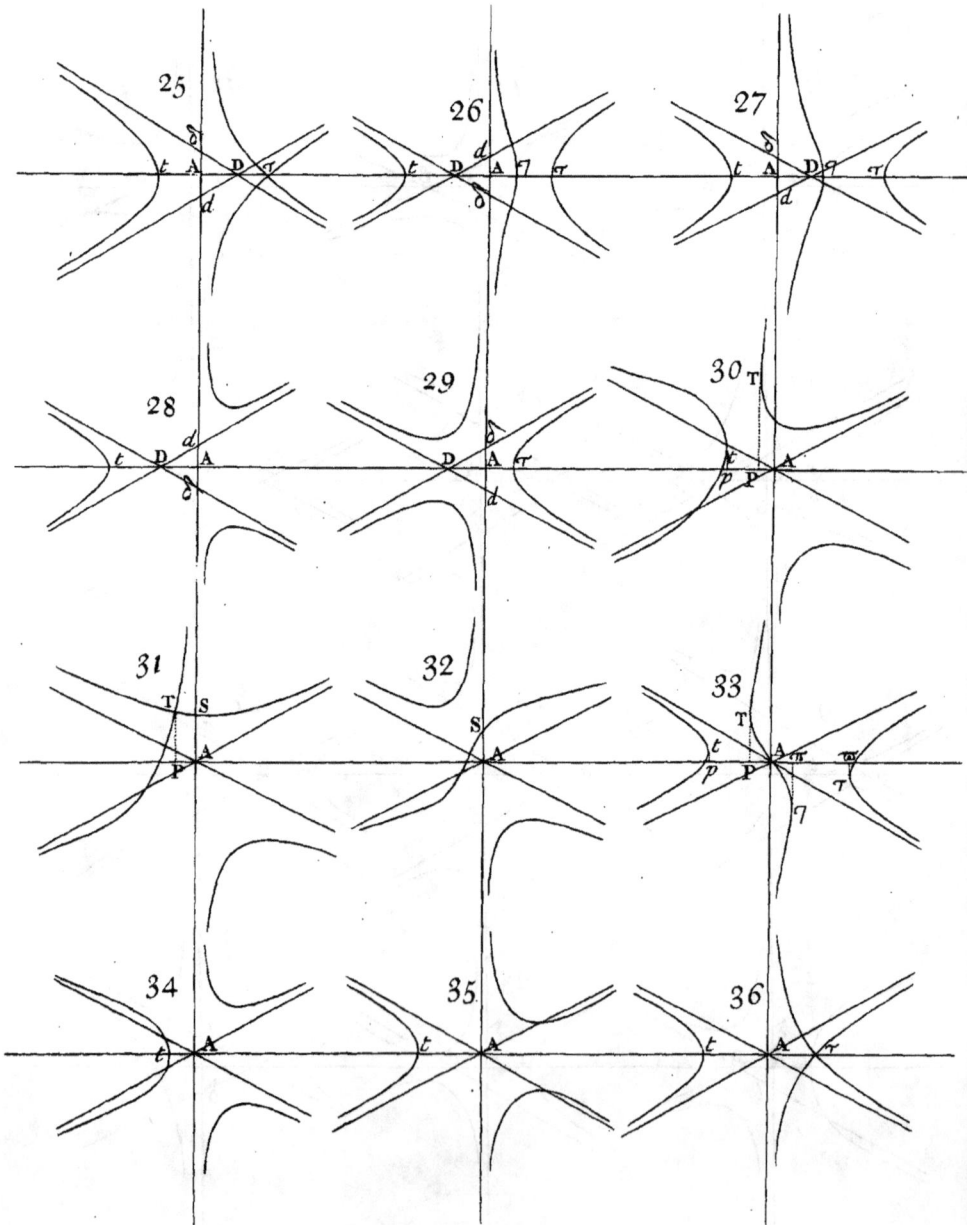

Curvarum Tab. III.

Si tres radices funt æquales, Nodus vertetur in *Fig. 41.*
cufpidem acutiffimum in vertice anguineæ. Et hæc
eft fpecies tricesima quinta.

Si e tribus radicibus ejufdem figni duæ maximæ *Fig. 43.*
A *p* & A *ϖ* fibi mutuo æquantur, Ovalis in *punctum*
evanuit. Quæ fpecies eft tricefima fexta.

Si radices duæ quævis imaginariæ funt, fola manebit Anguinea *pura* fine Ovali, decuffatione, cufpide vel puncto conjugato. Si Anguinea illa non *Fig. 42.*
tranfit per punctum A fpecies eft tricefima feptima,
fin tranfit per punctum illud A (id quod contingit *Fig. 43.*
ubi termini b ac d defunt,) punctum illud A erit
centrum figuræ rectas omnes per ipfum ductas &
ad Curvam utrinq; terminatas bifecans. Et hæc
eft fpecies tricefima octava.

In altero cafu ubi terminus e y deeft & propterea XX.
figura Diametrum habet, fi æquationis hujus a x³ *Hyperbolæ fep-*
= b x x + c x + d radices omnes A T, A t, A τ, funt *tem defectivæ, di-*
ametrum haben-
reales, inæquales & ejufdem figni, figura erit Hyper- *tes.*
bola Conchoidalis cum *Ovali* ad convexitatem. Quæ *Fig. 45.*
eft fpecies tricefima nona.

Si duæ radices funt inæquales & ejufdem figni & *Fig. 44.*
tertia eft figni contrarii, *Ovalis* jacebit ad concavitatem Conchoidalis. Eftq; fpecies quadragefima.

Si radices duæ minores A T, A t, funt æquales *Fig. 46.*
& tertia A τ eft ejufdem figni, Ovalis & Conchoidalis jungentur fefe decuffando in modum *Nodi.*
Quæ fpecies eft quadragefima prima.

Si tres radices funt æquales, Nodus mutabitur in *Fig. 47.*
Cufpidem & figura erit *Ciffois Veterum.* Et hæc eft
fpecies quadragefima fecunda.

Si

Fig. 49.   Si radices duæ majores funt æquales, & tertia eft ejufdem figni, Conchoidalis habebit *punctum* conjugatum ad convexitatem fuam, eftq; fpecies quadragefima tertia.

Fig. 49.   Si radices duæ funt æquales & tertia eft figni contrarii Conchoidalis habebit *punctum* conjugatum ad concavitatem fuam, eftq; fpecies quadragefima quarta.

Fig. 48,49.   Si radices duæ funt impoffibiles habebitur Conchoidalis *pura,* fine Ovali, Nodo, Cufpide vel puncto conjugato. Quæ fpecies eft quadragefima quinta.

XXI.
Hyperbola feptem Parabolicæ Diametrum non habentes.
     Siquando in primo æquationum cafu terminus $a x^3$ deeft & terminus $b x x$ non deeft, Figura erit Hyperbola Parabolica duo habens crura Hyperbolica ad unam Afymptoton SAG & duo Parabolica in plagam unam & eandem convergentia. Si terminus $e y$ non deeft figura nullam habebit diametrum, fin deeft habebit unicam. In priori cafu fpecies funt hæ.

Fig. 50.   Si tres radices A P, A ϖ, A π æquationis hujus $b x^3 + c x + d x + \frac{1}{4} e e = 0$ funt inæquales & ejufdem figni, figura conftabit ex *Ovali* & aliis duabus Curvis quæ partim Hyperbolicæ funt & partim Parabolicæ. Nempe crura Parabolica continuo ductu junguntur cruribus Hyperbolicis fibi proximis. Et hæc eft fpecies quadregefima fexta.

Fig. 51   Si radices duæ minores funt æquales & tertia eft ejufdem figni, Ovalis & una Curvarum illarum Hyperbolo-Parabolicarum junguntur & fe decuffant in formam *Nodi.* Quæ fpecies eft quadragefima feptima

Si

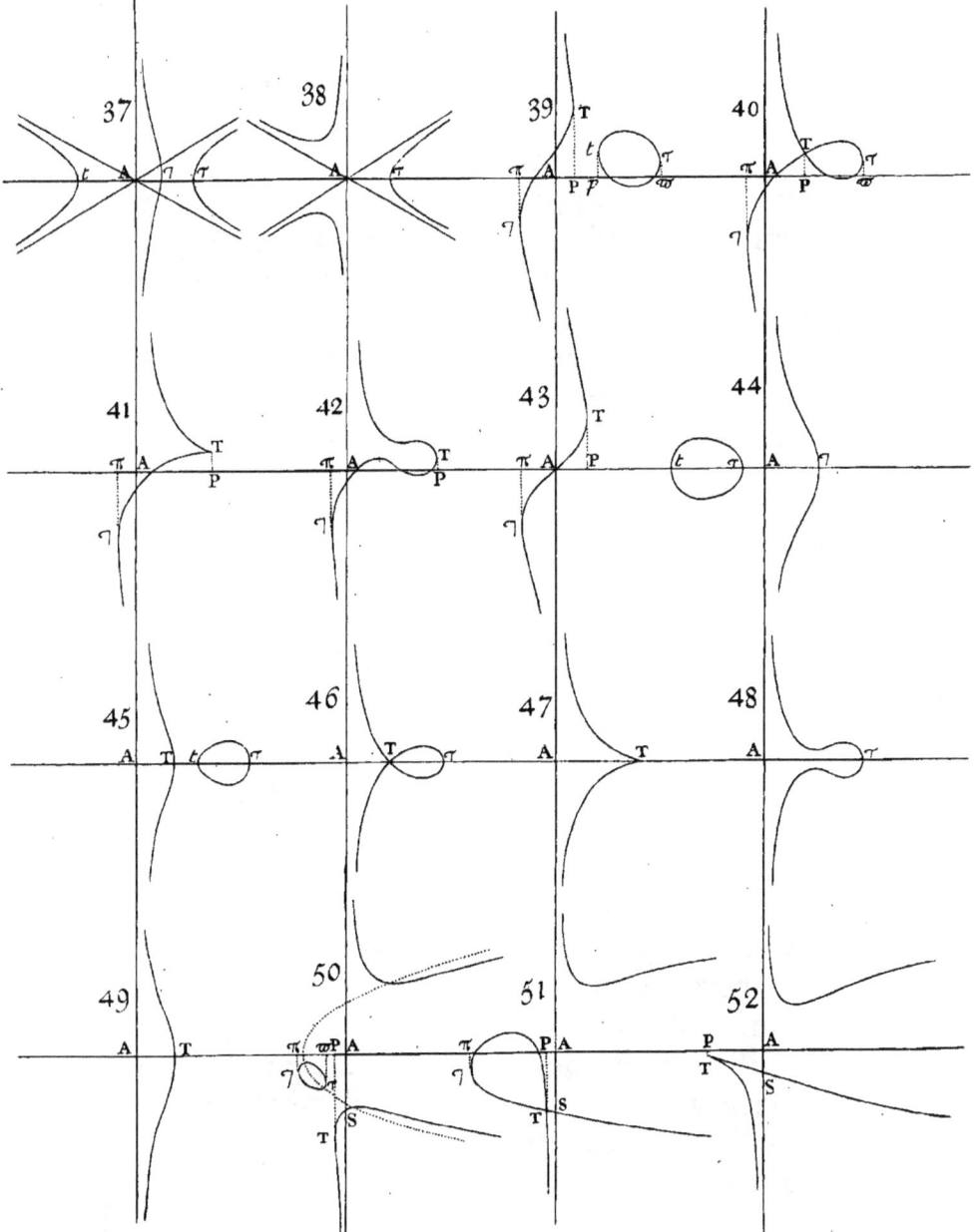

Curvarum Tab. IV.

Si tres radices funt æquales, Nodus ille in Cuf- *Fig. 52.*
pidem vertitur. Eftq; fpecies quadragefima octava.

Si radices duæ majores funt æquales & tertia eft *Fig. 53.*
ejufdem figni, Ovalis in *punctum* conjugatum evanuit. Quæ fpecies eft quadragefima nona.

Si duæ radices funt impoffibiles, manebunt *puræ Fig. 53,54.*
illæ duæ curvæ Hyperbolo-parabolicæ fine Ovali,
decuffatione, cufpide vel puncto conjugato, & fpeciem quinquagefimam conftituent.

Si radices duæ funt æquales & tertia eft figni con- *Fig. 55.*
trarii, Curvæ illæ hyperbolo-parabolicæ junguntur
fefe decuffando in morem crucis. Eftq; fpecies quinquagefima prima.

Si radices duæ funt inæquales & ejufdem figni & *Fig. 56.*
tertia eft figni contrarii, figura evadet Hyperbola
anguinea circa Afymptoton A G, cum Parabola conjugata. Et hæc eft fpecies quinquagefima fecunda.

In altero cafu ubi terminus e y deeft & figura   XXII.
Diametrum habet, fi duæ. radices æquationis hujus *Hyperbolæ qua-*
 tuor Parabolicæ
$bxx + cx + d = 0$ funt impoffibiles, duæ habentur *Diametrum ha-*
bentes.
figuræ hyperbolo-parabolicæ a Diametro A B hinc *Fig. 57.*
inde æqualiter diftantes. Quæ fpecies eft quinquagefima tertia.

Si æquationis illius radices duæ funt impoffibiles, *Fig. 58.*
Figuræ hyperbolo-parabolicæ junguntur fefe decuffantes in morem crucis, & fpeciem quinquagefimam quartam conftituunt.

Si radices illæ funt inæquales & ejufdem figni, ha- *Fig. 59.*
betur Hyperbola Conchoidalis cum Parabola ex
eodem latere Afymptoti. Eftq; fpecies quinquagefima quinta.

                        Si

Fig. 60.

Si radices illæ funt figni contrarii, habetur Con-choidalis cum Parabola ad alteras partes Afymptoti. Quæ fpecies eft quinquagefima fexta.

Siquando in primo æquationum cafu terminus uterq; a x³ & b x x deeft, figura erit Hyperbolifmus fectionis alicujus Conicæ. Hyperbolifmum figuræ voco cujus Ordinata prodit applicando contentum fub Ordinata figuræ illius & recta data ad Abfciffam com-munem. Hac ratione linea recta vertitur in hyper-bolam Conicam, & fectio omnis Conica vertitur in aliquam figurarum quas hic Hyperbolifmos fectio-num Conicarum voco. Nam æquatio ad figuras de quibus agimus, nempe x y y $+$ e y $=$ c x $+$ d, feu

$$y = \frac{e \pm \sqrt{ee + 4 dx + 4\ cxx}}{2\ x} \quad \text{generatur appli-}$$

cando contentum fub Ordinata fectionis Conicæ

$$\frac{e \pm \sqrt{ee + 4 dx + 4 cxx}}{2\ m} \ \& \ \text{recta data m ad curvarum}$$

Abfciffam communem x. Unde liquet quod figura genita Hyperbolifmus erit Hyperbolæ, Ellipfeos vel Parabolæ perinde ut terminus c x affirmativus eft vel negativus vel nullus.

Hyperbolifmus Hyperbolæ tres habet afymptotos quarum una eft Ordinata prima & principalis A d, alteræ duæ funt parallelæ Abfciffæ A B & ab eadem hinc inde æqualiter diftant. In Ordinata principali A d cape A d, A σ hinc inde æquales quantitati $\sqrt{c}$ & per puncta d ac σ age d g, $\delta \gamma$ Afymptotos Ab-fciffæ A B parallelas.

Ubi terminus e y non deeft figura nullam ha-bet diametrum. In hoc cafu fi æquationis hujus c x x $+$ d x $+\frac{1}{4}$ e e $=$ o radices duæ A P, A *p* funt reales

&

& inæquales (nam æquales esse nequeunt nisi figura *Fig. 61.* sit Conica sectio) figura constabit ex tribus Hyperbolis sibi oppositis quarum una jacet inter asymptotos parallelas & alteræ duæ jacent extra. Et hæc est species quinquagesima septima.

Si radices illæ duæ sunt impossibiles, habentur Hyperbolæ duæ oppositæ extra asymptotos parallelas & Anguinea hyperbolica intra easdem. Hæc figura duarum est specierum. Nam centrum non habet *Fig. 62.* ubi terminus d non deest ; sed si terminus ille deest *Fig. 63.* punctum A est ejus centrum. Prior species est quinquagesima octava, posterior quinquagesima nona.

Quod si terminus e y deest, figura constabit ex *Fig. 64.* tribus hyperbolis oppositis quarum una jacet inter asymptotos parallelas & alteræ duæ jacent extra ut in specie quinquagesima quarta, & præterea diametrum habet quæ est abscissa A B. Et hæc est species sexagesima.

Hyperbolismus Ellipseos per hanc æquationem de- **XXIV.** finitur x y y + e y = c x + d, & unicam habet asymp- *Tres Hyperbolismi Ellipseos.* toton quæ est Ordinata principalis A d. Si terminus *Fig. 65.* e y non deest, figura est Hyperbola anguinea sine diametro atq; etiam sine centro si terminus d non deest. Quæ species est sexagesima prima.

At si terminus d deest, figura habet centrum sine *Fig. 66.* diametro & centrum ejus est punctum A. Species vero est sexagesima secunda.

Et si terminus e y deest & terminus d non deest, *Fig. 67.* figura est Conchoidalis ad asymptoton A G, habetq; diametrum sine centro, & diameter ejus est Abscissa A B. Quæ species est sexagesima tertia.

X x 2

Hyper-

XXV.
Duo Hyperbo-
lifmi Parabola.

Fig. 68.

Fig. 69.

Hyperbolifmus Parabolæ per hanc æquationem definitur x y y + e y = d ; & duas habet afymptotos, Abfciffam A B & Ordinatam primam & principalem A G. Hyperbolæ vero in hac figura funt duæ, non in afymptotôn angulis oppofitis fed in angulis qui funt deinceps jacentes, idq; ad utrumq; latus abfciffæ A B, & vel fine diametro fi terminus e y habetur, vel cum diametro fi terminus ille deeft. Quæ duæ fpecies funt fexagefima quarta & fexagefima quinta.

XXVI.
Tridens.

Fig. 76.

In fecundo æquationum cafu habebatur æquatio x y = a x³ + b x x + c x + d. Et figura in hoc cafu habet quatuor crura infinita quorum duo funt hyperbolica circa afymptoton A G in contrarias partes tendentia & duo Parabolica convergentia & cum prioribus fpeciem Tridentis fere efformantia. Eftq; hæc Figura Parabola illa per quam Cartefius æquationes fex dimenfionum conftruxit. Hæc eft igitur fpecies fexagefima fexta.

XXVII.
Parabola quin-
que divergentes.

Fig. 70, 71.

In tertio cafu æquatio erat y y = a x³ + b x x + c x + d, & Parabolum defignat cujus crura divergunt ab invicem & in contrarias partes infinite progrediuntur. Abfciffa A B eft ejus diameter & fpecies ejus funt quinq; fequentes.

Si æquationis a x³ + b x² + c x + d = o radices omnes A𝜏, A T, A t funt reales & inæquales, figura eft Parabola divergens campaniformis cum *Ovali* ad verticem. Et fpecies eft fexagefima feptima.

Fig. 72.
Fig. 73.

Si radices duæ funt æquales, Parabola prodit vel *nodata* contingendo Ovalem, vel *punctata* ob Ovalem infinite parvam. Quæ duæ fpecies funt fexagefima octava & fexagefima nona.

Si

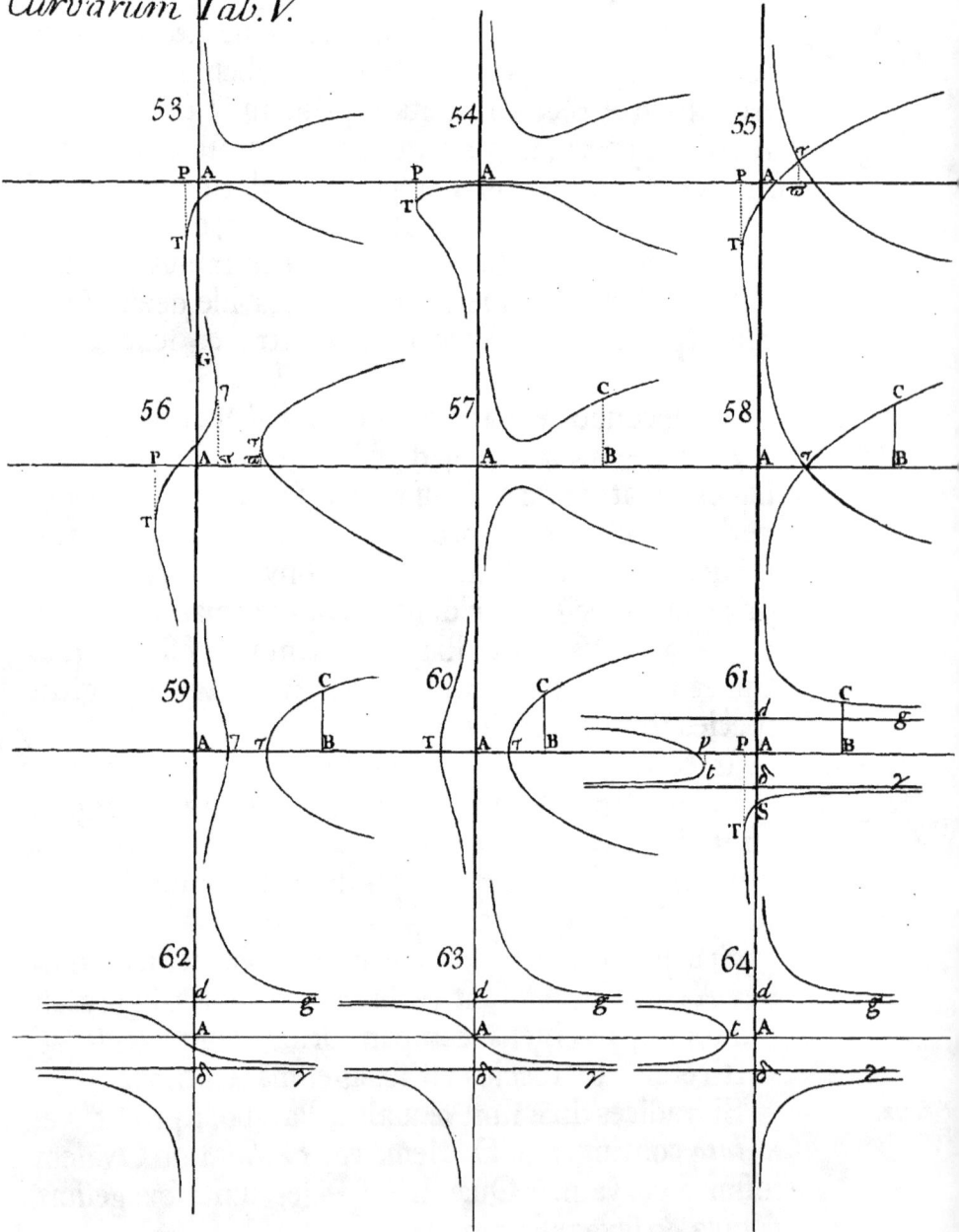

Curvarum Tab. V.

Si tres radices funt æquales Parabola erit *cufpi-* data in vertice.   Et hæc eft Parabola Neiliana quæ vulgo femicubica dicitur.

Fig. 75.

Si radices duæ funt impoffibiles, habetur Parabola *pura* campaniformis fpeciem feptuagefimam primam conftituens.

Fig. 73, 74.

In quarto cafu æquato erat $y = ax + bxx + cx + d$, & hæc æquatio Parabolam illam *Wallifianam* defignat quæ crura habet contraria & *cubica* dici folet.   Et fic fpecies omnino funt feptuaginta duæ.

XXVIII.
Parabola cubica.
Fig. 77.

Si in planum infinitum a punɛto lucido illumina-tum umbræ figurarum projiciantur, umbræ feɛtio-num Conicarum femper erunt feɛtiones Conicæ, eæ Curvarum fecundi generis femper erunt Curvæ fe-cundi generis, eæ curvarum tertii generis femper erunt Curvæ tertii generis, & fic deinceps in infini-tum.   Et quemadmodum Circulus umbram proji-ciendo generat feɛtiones omnes conicas, fic Parabolæ quinq; divergentes umbris fuis generant & exhi-bent alias omnes fecundi generis curvas , & fic Curvæ quædam fimpliciores aliorum generum inve-niri poffunt quæ alias omnes eorundem generum curvas umbris fuis a punɛto lucido in planum pro-jeɛtis formabunt.

XXIX.
Genefis Curva-rum per Umbras.

Diximus Curvas fecundi generis a linea reɛta in punɛtis tribus fecari poffe.   Horum duo nonnun-quam coincidunt.   Ut cum reɛta per Ovalem infi-nite parvam tranfit vel per concurfum duarum par-tium Curvæ fe mutuo fecantium vel in cufpidem coeuntium ducitur.   Et fiquando reɛtæ omnes in plagam

XXX.
Curvarum pun-ɛta duplicia.

plagam cruris alicujus infiniti tendentes Curvam
in unico tantum puncto fecant (ut fit in ordinatis
Parabolæ Cartefianæ & Parabolæ cubicæ, nec non in
rectis Abfciffæ Hyperbolifmorum Hyberbolæ & Parabolæ parallelis) concipiendum eft quod rectæ illæ
per alia duo Curvæ puncta ad infinitam diftantiam fita (ut ita dicam) tranfeunt. Hujufmodi
interfectiones duas coincidentes five ad finitam
fint diftantiam five ad infinitam, vocabimus punctum duplex.   Curvæ autem quæ habent punctum duplex defcribi poffunt per fequentia Theoremata.

XXXI.
*Theoremata de
Curvarum de-
fcriptione orga-
nica.*
Fig. 78.

1. Si anguli duo magnitudine dati PAD, PBD circa
polos pofitione datos A, B rotentur, & eorum crura
A P, B P concurfu fuo P percurrant lineam rectam ;
crura duo reliqua A D, B D concurfu fuo D defcribent fectionem Conicam per polos A, B tranfeuntem : præterquam ubi linea illa recta tranfit per polorum alterutrum A vel B, vel anguli B A D, A B D
fimul evanefcunt, quibus in cafibus punctum D defcribet lineam rectam.

2. Si crura prima A P, B P concurfu fuo P
percurrant fectionem Conicam per polum alterutrum A tranfeuntem, crura duo reliqua A D, B D
concurfu fuo D defcribent Curvam fecundi generis per polum alterum B tranfeuntem & punctum duplex habentem in polo primo A per quem
fectio Conica tranfit : præterquam ubi anguli
B A D, A B D fimul evanefcunt, quo cafu punctum

ctum D defcribet aliam fectionem Conicam per polum A tranfeuntem.

3. At fi fectio Conica quam punctum P percurrit tranfeat per neutrum polorum A, B, punctum D defcribet curvam fecundi vel tertii generis punctum duplex habentem. Et punctum illud duplex in concurfu crurum defcribentium, A D, B D invenietur ubi anguli B A P, A B P fimul evanefcunt. Curva autem defcripta fecundi erit generis fi anguli B A D, A B D fimul evanefcunt, alias erit tertii generis & alia duo habebit puncta duplicia in polis A & B.

Jam fectio Conica determinatur ex datis ejus punctis quinq; & per eadem fic defcribi poteft. Dentur ejus puncta quinq; A, B, C, D, E. Jungantur eorum tria quævis A, B, C & trianguli A B C rotentur anguli duo quivis C A B, C B A circa vertices fuos A & B, & ubi crurum A C, B C interfectio C fucceffive applicatur ad puncta duo reliqua D, E, incidat interfectio crurum reliquorum A B & B A in puncta P & Q. Agatur & infinite producatur recta P Q, & anguli mobiles ita rotentur ut interfectio crurum A B, B A percurrat rectam P Q, & crurum reliquorum interfectio C defcribet propofitam fectionem Conicam per Theorema primum.

XXXII.
Sectionum Conicarum defcriptio per data quinque puncta.

Curvæ omnes fecundi generis punctum duplex habentes determinantur ex datis earum punctis feptem, quorum unum eft punctum illud duplex, &

XXXIII.
Curvarum fecundi generis punctum duplex habentium defcriptio per data feptem puncta.

& per eadem puncta sic describi possunt. Dentur
Curvæ describendæ puncta quælibet septem A, B, C,
D, E, F, G quorum A est punctum duplex. Jungantur punctum A & alia duo quævis e punctis puta
B & C; & trianguli A B C rotetur tum angulus
C A B circa verticem suum A, tum angulorum reliquorum alteruter A·B C circa verticem suum B. Et
ubi crurum A C, B C concursus C successive applicatur ad puncta quatuor reliqua D, E, F, G incidat
concursus crurum reliquorum A B & B A in puncta
quatuor P, Q, R, S. Per puncta illa quatuor &
quintum A describatur sectio Conica, & anguli præ-
fati C A B, C B A ita rotentur ut crurum A B, B A
concursus percurrat sectionem illam Conicam, &
concursus reliquorum crurum A C, B C describet
Curvam propositam per Theorema secundum.

Si vice puncti C datur positione recta B C quæ
Curvam describendam tangit in B, lineæ A D, A P
coincident, & vice anguli D A P habebitur linea recta
circa polum A rotanda.

Si punctum duplex A infinite distat debebit Recta
ad plagam puncti illius perpetuo dirigi & motu pa-
rallelo ferri interea dum angulus A B C circa polum
B rotatur.

Describi etiam possunt hæ curvæ paulo aliter per
Theorema tertium, sed descriptionem simpliciorem
posuisse sufficit.

Eadem methodo Curvas tertii, quarti & superiorum generum describere licet, non omnes quidem
sed quotquot ratione aliqua commoda per motum
localem describi possunt. Nam curvam aliquam
secundi

fecundi vel fuperioris generis punctum duplex non habentem commode defcribere Problema eft inter difficiliora numerandum.

Curvarum ufus in Geometria eft ut per earum interfectiones Problemata folvantur. Proponatur æquatio conftruenda dimenfionum novem $x^9 * + b x^7$ $+ c x^6 + d x^5 + e x^4 + f x^3 + g x x + h x + k = o$. Ubi $+ m$

XXXIV. *Conftructio æquationum per defcriptionem Curvarum.*

b, c, d, &c. fignificant quantitates quafvis datas fignis fuis $+$ & $-$ affectas. Affumatur æquatio ad Parabolam cubicam $x^3 = y$, & æquatio prior, fcribendo y pro x^3, evadet $y^3 + b x y y + c y y + d x x y$ $+ e x y + m y + f x^3 + g x x + h x + k = o$, æquatio ad Curvam aliam fecundi generis. Ubi m vel f deeffe poteft vel pro lubitu affumi. Et per harum Curvarum defcriptiones & interfectiones dabuntur radices æquationis conftruendæ. Parabolam cubicam femel defcribere fufficit.

Si æquatio conftruenda per defectum duorum terminorum ultimorum h x & k reducatur ad feptem dimenfiones, Curva altera delendo m, habebit punctum duplex in principio abfciffæ, & inde facile defcribi poteft ut fupra.

Si æquatio conftruenda per defectum terminorum trium ultimorum $g x x + h x + k$ reducatur ad fex dimenfiones, Curva altera delendo f evadet fectio Conica.

Et fi per defectum fex ultimorum terminorum æquatio conftruenda reducatur ad tres dimenfiones, incidetur in conftructionem *Wallifianam* per Parabolam cubicam & lineam rectam.

Y y

Con-

Conftrui etiam poffunt æquationes per Hyperbo-
lifimum Parabolæ cum diametro. Ut fi conftruenda
fit hæc æquatio dimenfionum novem termino penul-
timo carens, $a + cxx + dx^3 + ex^4 + fx + gx^6 + hx^7$
$$+ m$$
$+ kx^8 + lx^9 = 0$; affumatur æquatio ad Hyperbolifmum illum $xxy = 1$, & fcribendo y pro $\frac{1}{xx}$, æquatio
conftruenda vertetur in hanc $ay^3 + cyy + dxyy + ey$
$+ fxy + mxxy + g + hx + kxx + lx^3 = 0$, quæ curvam fecundi generis defignat cujus defcriptione
Problema folvetur. Et quantitatum m ac g alterutra hic deeffe poteft, vel pro lubitu affumi.

Per Parabolam cubicam & Curvas tertii generis
conftruuntur etiam æquationes omnes dimenfionum
non plufquam duodecim, & per eandem Parabolam
& curvas quarti generis conftruuntur omnes dimenfionum non plufquam quindecim, Et fic deinceps in
infinitum. Et curvæ illæ tertii quarti & fuperiorum
generum defcribi femper poffunt inveniendo eorum
puncta per Geometriam planam. Ut fi conftruenda
fit æquatio $x^{11} * + ax^{10} + bx^9 + cx^8 + dx^7 + ex^6 + fx^5$
$+ gx^4 + hx^3 + ixx + kx + l = 0$, & defcripta
habeatur Parabola Cubica; fit æquatio ad Parabolam illam cubicam $x^3 = y$, & fcribendo y
pro x^3 æquatio conftruenda vertetur in hanc
$y^4 + axy^3 + cxxyy + fxxy + ixx = 0$, quæ eft
$+ b + dx + gx + kx$
$+ e + h + l$
æquatio ad Curvam tertii generis cujus defcriptione
Problema folvetur. Defcribi autem poteft hæc Curva
inveniendo ejus puncta per Geometriam planam,prop-
terea quod indeterminata quantitas x non nifi ad
duas dimenfiones afcendit.

Curvarum Tab. VI.

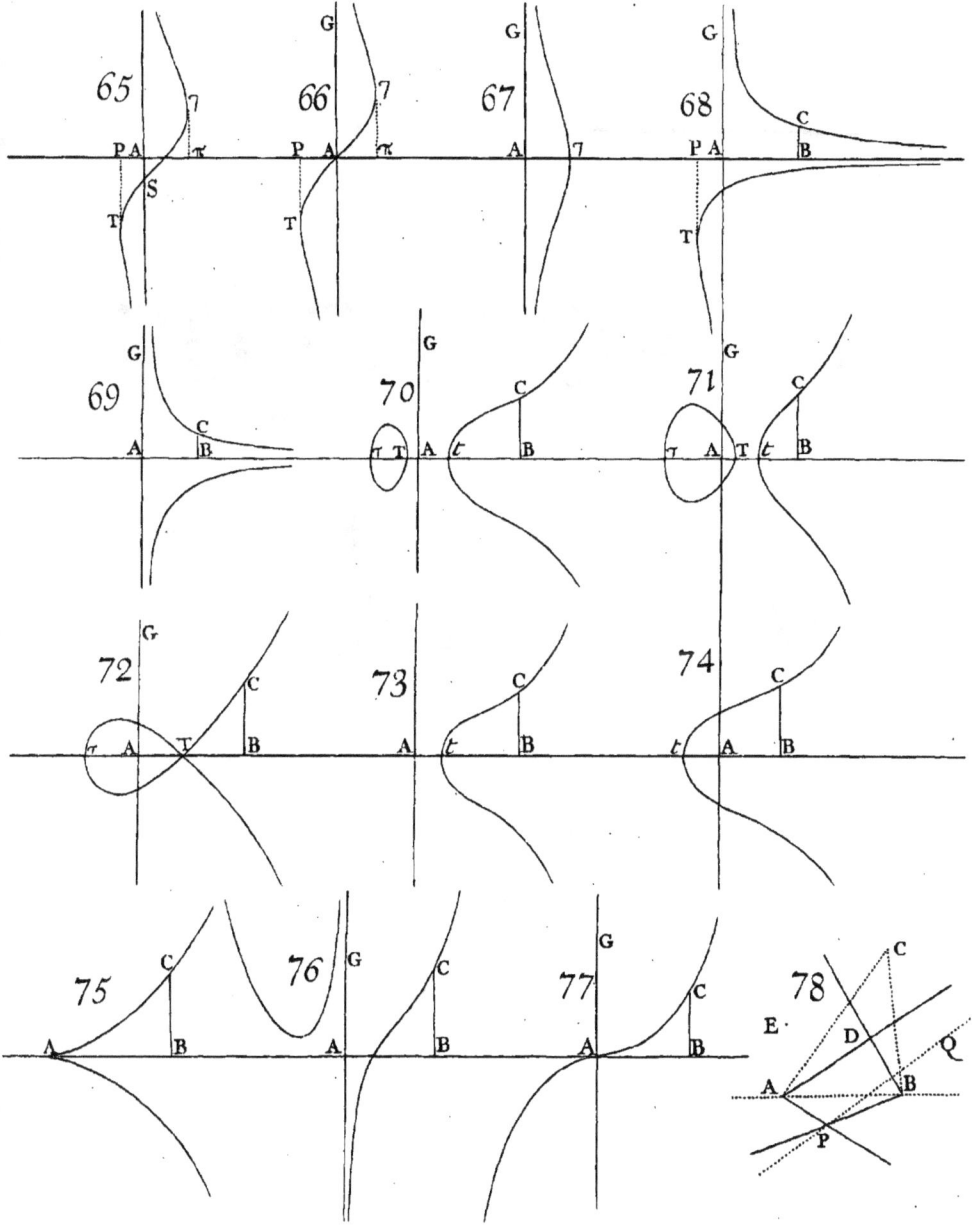

65   66   67   68

69   70   71

72   73   74

75   76   77   78

TRACTATUS

D E

Quadratura Curvarum.

INTRODUCTIO.

Quantitates Mathematicas non ut ex partibus quam minimis conftantes, fed ut motu continuo defcriptas hic confidero. Lineæ defcribuntur ac defcribendo generantur non per appofitionem partium fed per motum continuum punctorum, fuperficies per motum linearum, folida per motum fuperficierum, anguli per rotationem laterum, tempora per fluxum continuum, & fic in cæteris. Hæ Genefes in rerum natura locum vere habent & in motu corporum quotidie cernuntur. Et ad hunc modum Veteres ducendo rectas mobiles in longitudinem rectarum immobilium genefin docuerunt rectangulorum.

Confiderando igitur quod quantitates æqualibus temporibus crefcentes & crefcendo genitæ, pro velocitate majori vel minori qua crefcunt ac generantur, evadunt majores vel minores ; methodum quærebam

deter-

determinandi quantitates ex velocitalibus motuum vel incrementorum quibus generantur ; & has motuum vel incrementorum velocitates nominando *Fluxiones* & quantitates genitas nominando *Fluentes,* incidi paulatim *Annis* 1665 & 1666 in Methodum Fluxionum qua hic ufus fum in Quadratura Curvarum.

Fluxiones funt quam proxime ut Fluentium augmenta æqualibus temporis particulis quam minimis genita, & ut accurate loquar, funt in prima ratione augmentorum nafcentium ; exponi autem poffunt per lineas quafcunq; quæ funt ipfis proportionales. Ut

Fig. 1.

fi areæ A B C , A B D G Ordinatis B C , B D fuper bafi A B uniformi cum motu progredientibus defcribantur, harum arearum fluxiones erunt inter fe ut Ordinatæ defcribentes B C & B D, & per Ordinatas illas exponi poffunt, propterea quod Ordinatæ illæ funt ut arearum augmenta nafcentia. Progrediatur Ordinata B C de loco fuo B C in locum quemvis novum b c. Compleatur parallelogrammum B C E b, ac ducatur recta V T H quæ Curvam tangat in C ipfifq; b c & B A productis occurrat in T & V : & Abfciffæ A B, Ordinatæ B C, & Lineæ Curvæ A C c augmenta modo genita erunt B b, E c & C c ; & in horum augmentorum nafcentium ratione prima funt latera trianguli C E T, ideoq; fluxiones ipfarum A B, B C & A C funt ut trianguli illius C E T latera C E, E T & C T & per eadem latera exponi poffunt, vel quod perinde eft per latera trianguli confimilis V B C.

Eodem recidit fi fumantur fluxiones in ultima ratione partium evanefcentium. Agatur recta C c & producatur eadem ad K. Redeat Ordinata b c

in

in locum fuum priorem B C, & coeuntibus punctis
C & c, recta C K coincidet cum tangente CH, &
triangulum evanefcens CEc in ultima fua forma
evadet fimile triangulo CET, & ejus latera evanef-
centia CE, Ec & Cc erunt ultimo inter fe ut funt
trianguli alterius CET latera CE, ET & CT, &
propterea in hac ratione funt fluxiones linearum AB,
BC & AC. Si puncta C & c parvo quovis intervallo ab invicem diftant recta CK parvo intervallo a
tangente CH diftabit. Ut recta CK cum tangente
CH coincidat & rationes ultimæ linearum CE, Ec &
Cc inveniantur, debent puncta C & c coire & omnino coincidere. Errores quam minimi in rebus
mathematicis non funt contemnendi.

Simili argumento fi circulus centro B radio BC
defcriptus in longitudinem Abfciffæ AB ad angulos
rectos uniformi cum motu ducatur, fluxio folidi ge-
niti ABC erit ut circulus ille generans, & fluxio fu-
perficiei ejus erit ut perimeter Circuli illius &
fluxio lineæ curvæ AC conjunctim. Nam quo tempore folidum ABC generatur ducendo circulum
illum in longitudinem Abfciffæ AB, eodem fuperficies ejus generatur ducendo perimetrum circuli illius in longitudinem Curvæ AC.

Recta PB circa polum datum P revolvens fecet aliam Fig. 2.
pofitione datam rectam AB: quæritur proportio fluxio-
num rectarum illarum AB & PB. Progrediatur
recta PB de loco fuo PB in locum novum Pb. In
Pb capiatur PC ipfi PB æqualis, & ad AB ducatur
PD fic, ut angulus bPD æqualis fit angulo bBC;
& ob fimilitudinem triangulorum bBC, bPD erit
augmentum Bb ad augmentum Cb ut Pb ad Db.

Redeat

Redeat jam P b in locum fuum priorem P B ut augmenta illa evanefcant, & evanefcentium ratio ultima, id eft ratio ultima P b ad D b, ea erit quæ eft
P B ad D B, exiftente angulo P D B recto, & propterea in hac ratione eft fluxio ipfius A B ad fluxionem
ipfius P B.

Fig. 3.

*Recta P B circa datum Polum P revolvens fecet
alias duas pofitione datas rectas A B & A E in B &
E : quæritur proportio fluxionum rectarum illarum
A B & A E.* Progrediatur recta revolvens P B de
loco fuo P B in locum novum P b rectas A B, A E in
punctis b & e fecantem, & rectæ A E parallela B C
ducatur ipfi P b occurrens in C, & erit B b ad B C ut
A b ad A e, & B C ad E e ut P B ad P E, & conjunctis
rationibus B b ad E e ut A b×P B ad A e×P E.
Redeat jam linea P b in locum fuum priorem P B, &
augmentum evanefcens B b erit ad augmentum evanefcens E e ut A B×P B ad A E×P E, ideoq; in
hac ratione eft fluxio rectæ A B ad fluxionem rectæ
A E.

Hinc fi recta revolvens P B lineas quafvis Curvas
pofitione datas fecet in punctis B & E, & rectæ jam
mobiles A B, A E Curvas illas tangant in Sectionum
punctis B & E : erit fluxio Curvæ quam recta, A B
tangit ad fluxionem Curvæ quam recta A E tangit
ut A B×P B ad A E×P E. Id quod etiam eveniet
fi recta P B Curvam aliquam pofitione datam perpetuo tangat in puncto mobili P.

*Fluat quantitas x uniformiter & invenienda fit fluxio
quantitatis xⁿ.* Quo tempore quantitas x fluendo
evadit x + o, quantitas x^n evadet $\overline{x + o}|^n$, id eft
per methodum ferierum infinitarum, $x^n {+} n o x^{n-1}$

+

$+\frac{nn-n}{2}oox^{n-2}+\&c.$ Et augmenta o & $nox^{n-1}+\frac{nn-n}{2}oox^{n-2}$ $+\&c.$ funt ad invicem ut 1 & $nx^{n-1}+\frac{nn-n}{2}ox^{n-2}+\&c.$ Evanefcant jam augmenta illa, & eorum ratio ultima erit 1 ad nx^{n-1} : ideoq; fluxio quantitatis x eft ad fluxionem quantitatis x^n ut 1 ad nx^{n-1}.

Similibus argumentis per methodum rationum primarum & ultimarum colligi poffunt fluxiones linearum feu rectarum feu curvarum in cafibus quibufcunque, ut & fluxiones fuperficierum, angulorum & aliarum quantitatum. In finitis autem quantitatibus Analyfin fic inftituere, & finitarum nafcentium vel evanefcentium rationes primas vel ultimas inveftigare, confonum eft Geometriæ Veterum : & volui oftendere quod in Methodo Fluxionum non opus fit figuras infinite parvas in Geometriam introducere. Peragi tamen poteft Analyfis in figuris quibufcunq; feu finitis feu infinite parvis quæ figuris evanefcentibus finguntur fimiles, ut & in figuris quæ pro infinite parvis haberi folent, modo caute procedas.

Ex Fluxionibus invenire Fluentes Problema difficilius eft, & folutionis primus gradus æquipollet Quadraturæ Curvarum ; de qua fequentia olim fcripfi.

D E

TRACTATUS

DE

Quadratura Curvarum.

Quantitates indeterminatas ut motu perpetuo crescentes vel decrescentes, id est ut fluentes vel defluentes in sequentibus considero, designoq; literis z, y, x, v, & earum fluxiones seu celeritates crescendi noto iisdem literis punctatis $\dot{z}$, $\dot{y}$, $\dot{x}$, $\dot{v}$. Sunt & harum fluxionum fluxiones seu mutationes magis aut minus celeres quas ipsarum z, y, x, v fluxiones secundas nominare licet & sic dignare $\ddot{z}$, $\ddot{y}$, $\ddot{x}$, $\ddot{v}$, & harum fluxiones primas seu ipsarum z, y, x, v fluxiones tertias sic $\dddot{z}$, $\dddot{y}$, $\dddot{x}$, $\dddot{v}$, & quartas sic $\ddddot{z}$, $\ddddot{y}$, $\ddddot{x}$, $\ddddot{v}$. Et quemadmodum $\dddot{z}$, $\dddot{y}$, $\dddot{x}$, $\dddot{v}$ sunt fluxiones quantitatum $\ddot{z}$, $\ddot{y}$, $\ddot{x}$, $\ddot{v}$, & hæ sunt fluxiones quantitatum $\dot{z}$, $\dot{y}$, $\dot{x}$, $\dot{v}$ & hæ sunt fluxiones quantitatum primarum z, y, x, v : sic hæ quantitates considerari possunt ut fluxiones aliarum quas sic designabo,

z,

$\dot{z}, \dot{y}, \dot{x}, \dot{v}$, & hæ ut fluxiones aliarum $\ddot{z}, \ddot{y}, \ddot{x}, \ddot{v}$, & hæ ut fluxiones aliarum $\dddot{z}, \dddot{y}, \dddot{x}, \dddot{v}$. Defignant igitur $\ddot{z}, \dot{z}, z, \ddot{z}, \ddot{z}, \ddot{z}, \dddot{z}, z$ &c. feriem quantitatum quarum quælibet pofterior eft fluxio præcedentis & quælibet prior eft fluens quantitas fluxionem habens fubfequentem. Similis eft feries $\overline{\sqrt{az-zz}}''$, $\overline{\sqrt{az-zz}}'$,

$\overline{\sqrt{az-zz}}$, $\overline{\sqrt{az-zz}}^{\cdot}$, $\overline{\sqrt{az-zz}}^{\cdot\cdot}$, $\overline{\sqrt{az-zz}}^{\cdot\cdot\cdot}$, ut &

feries $\overline{\dfrac{az+z^2}{a-z}}''$, $\overline{\dfrac{az+z^2}{a-z}}'$, $\overline{\dfrac{az+z^2}{a-z}}$, $\overline{\dfrac{az+z^2}{a-z}}^{\cdot}$, $\overline{\dfrac{az+z^2}{a-z}}^{\cdot\cdot}$,

$\overline{\dfrac{az+z^2}{a-z}}^{\,9}_{\cdot\cdot\cdot}$ Et notandum eft quod quantitas quælibet

prior in his feriebus eft ut area figuræ curviliniæ cujus ordinatim applicata rectangula eft quantitas

pofterior & abfciffa eft z : uti $\overline{\sqrt{az-zz}}'$ area curvæ

cujus ordinata eft $\sqrt{az-zz}$ & abfciffa z. Quo autem fpectant hæc omnia patebit in Propofitionibus quæ fequuntur.

Z z 2   PROP.

PROP. I. PROB. I.

Data æquatione quotcunq; fluentes quantitates involvente, invenire fluxiones.

Solutio.

Multiplicetur omnis æquationis terminus per indicem dignitatis quantitatis cujufq; fluentis quam involvit, & in fingulis multiplicationibus mutetur dignitatis latus in fluxionem fuam, & aggregatum factorum omnium fub propriis fignis erit æquatio nova.

Explicatio.

Sunto a, b, c, d &c. quantitates determinatæ & immutabiles, & proponatur æquatio quævis quantitates fluentes z, y, x &c. involvens, uti $x^3 - xyy + aaz - b^3 = 0$. Multiplicentur termini primo per indices dignitatum x, & in fingulis multiplicationibus pro dignitatis latere, feu x unius dimenfionis, fcribatur $\dot{x}$, & fumma factorum erit $3\dot{x}x^2 - \dot{x}yy$. Idem fiat in y & prodibit $-2\dot{y}xy$. Idem fiat in z & prodibit $aa\dot{z}$. Ponatur fumma factorum æqualis nihilo, & habebitur æquatio $3\dot{x}x^2 - \dot{x}yy - 2\dot{y}xy + aa\dot{z} = 0$. Dico quod hac æquatione definitur relatio fluxionum.

De-

Demonstratio.

Nam fit o quantitas admodum parva & funto oż, oẏ, oẋ, quantitatum z, y, x momenta id est incrementa momentanea fynchrona. Et fi quantitates fluentes jam funt z, y & x, hæ poft momentum temporis incrementis fuis oż, oẏ, oẋ auctæ, evadent z+oż, y+oẏ, x+oẋ, quæ in æquatione prima pro z, y & x fcriptæ dant æquationem x³+3xxoẋ +3xooẍx+o³ẋ³—xyy—oxẏẏ—2xoyẏ—2xooyẏ —xooẏẏ—xo³ẏẏ+aaz+aaoż—b3=o. Subducatur æquatio prior, & refiduum divifum per o erit 3xxẋ² +3xxoẍx+ẋ³oo—xẏẏ—2xẏẏ—2xoyẏ—xoẏẏ—xooẏẏ +aaż=o. Minuatur quantitas o in infinitum,& neglectis terminis evanefcentibus reftabit 3xẋ²—xẏẏ —2xyẏ+aaż=o. Q. E. D.

Explicatio plenior.

Ad eundem modum fi æquatio effet x³—xyy +aa √‾a‾x‾—‾y‾y—b3=o, produceretur 3x²ẋ—xẏẏ —2xyẏ+aa√‾a‾x‾—‾y‾y=o. Ubi fi fluxionem √‾a‾x‾—‾y‾y tollere velis, pone √‾a‾x‾—‾y‾y=z, & erit ax—yy=z²,

&

& (per hanc Propofitionem) $\dot{a}x - {}^2\dot{y}y = {}^2\dot{z}z$ feu

$\dfrac{\dot{a}x - 2\dot{y}y}{{}^2z} = \dot{z}$, hoc eft $\dfrac{\dot{a}x - 2\dot{y}y}{{}^2\sqrt{ax - yy}} = \sqrt{ax - yy}$. Et

inde $3x^2\dot{x} - \dot{x}yy - 2x\dot{y}y + \dfrac{a^3\dot{x} - 2aa\dot{y}y}{{}^2\sqrt{ax - yy}} = 0$

Et per operationem repetitam pergitur ad fluxiones fecundas, tertias & fequentes. Sit æquatio $zy^3 - z^4 + a^4 = 0$, & fiet per operationem primam

$\dot{z}y^3 + 3z\dot{y}y^2 - 4\dot{z}z^3 = 0$, per fecundam $\ddot{z}y^3 + 6\dot{z}\dot{y}y^2$

$+ 3z\ddot{y}y^2 + 6z\dot{y}^2y - 4\ddot{z}z^3 - 12\dot{z}^2z^2 = 0$, per tertiam

$\dddot{z}y^3 + 9\ddot{z}\dot{y}y^2 + 9\dot{z}\ddot{y}y^2 + 18\dot{z}\dot{y}^2y + 3z\dddot{y}y^2 + 18z\ddot{y}\dot{y}y$

$+ 6z\dot{y}^3 - 4\dddot{z}z^3 - 36\ddot{z}\dot{z}z^2 - 24\dot{z}^3z = 0.$

Ubi vero fic pergitur ad fluxiones fecundas, tertias & fequentes, convenit quantitatem aliquam ut uniformiter fluentem confiderare,& pro ejus fluxione prima unitatem fcribere, pro fecunda vero & fequentibus nihil. Sit æquatio $zy^3 - z^4 + a^4 = 0$, ut fupra; & fluat z uniformiter, fitq; ejus fluxio unitas, & fiet per operationem primam $y^3 + 3z\dot{y}y^2 - 4z^3 = 0,$

per fecundam $6\dot{y}y^2 + 3z\ddot{y}y^2 + 6z\dot{y}^2y - 12z^2 = 0,$

per tertiam $9\ddot{y}y^2 + 18\dot{y}^2y + 3z\dddot{y}y^2 + 18z\ddot{y}\dot{y}y + 6zy^3$ $- 24z = 0.$

In

In hujus autem generis æquationibus concipiendum eft quod fluxiones in fingulis terminis fint ejufdem ordinis, id eft vel omnes primi ordinis $\dot{y}$, $\dot{z}$,
vel omnes fecundi $\ddot{y}$, $\dot{y}^2$, $\dot{y}\dot{z}$, $\dot{z}^2$, vel omnes tertii
$\dddot{y}$, $\ddot{y}\dot{y}$, $\ddot{y}\dot{z}$, $\dot{y}^3$, $\dot{y}^2\dot{z}$, $\dot{y}\dot{z}^2$ $\dot{z}^3$ &c. Et ubi res aliter fe
habet complendus eft ordo per fubintelleëtas fluxio-
nes quantitatis uniformiter fluentis. Sic æquatio
noviffima complendo ordinem tertium fit $9\dot{z}\ddot{y}y^2$
$+18\dot{z}\dot{y}^2y+3\ddot{z}\dot{y}y^2+18\ddot{z}\dot{y}yy+6\dot{z}y^3-24z\dot{z}z^3=0.$

PROP. II. PROB. II.

Invenire Curvas quæ quadrari poffunt.

Sit A B C figura invenienda, BC Ordinatim ap- *Fig.* 4.
plicata reëtangula, & AB abfciffa. Producatur
CB ad E ut fit BE=1, & compleatur parallelogrammum ABED: & arearum ABC, ABED
fluxiones erunt ut BC & BE. Affumatur igitur
æquatio quævis qua relatio arearum definiatur, &
inde dabitur relatio ordinatarum BC & BE per
Prop. I. Q. E. I.

Hujus rei exempla habentur in Propofitionibus
duabus fequentibus.

PROP.

PROP. III. THEOR. I.

Si pro abſciſſa A B & area A E ſeu AB×1 pro-
miſcue ſcribatur z, & ſi pro e $-|-$fz^n $+$gz^{2n} $-|-$hz$^{3n}-|-$&c.
ſcribatur R: fit autem area Curvæ z$^\theta$R$^\lambda$ erit.
ordinatim applicata BC=

$$\theta e \overset{+}{+} \overset{\theta}{_{\lambda n}} fz^n \overset{+}{+} \overset{\theta}{_{2\lambda n}} gz^{2n} \overset{+}{+} \overset{\theta}{_{3\lambda n}} hz^{3n} + \&c. \text{ in } z^{\theta-1} R^{\lambda-1}.$$

Demonſtratio.

Nam ſi fit z$^\theta$R$^\lambda$=v, erit per Prop. 1, $\theta zz^{\theta-1}R_\lambda$
$-|-\lambda z^\theta \dot{R}R^{\lambda-1} = \dot{v}$. Pro R$^\lambda$ in primo æquationis termino & z$^\theta$ in ſecundo ſcribe RR$^{\lambda-1}$ & zz$^{\theta-1}$, & fiet
$\theta zR + \lambda z\dot{R}$ in $z^{\theta-1} R^\lambda_{-1} = \dot{v}$. Erat autem R=e $-|-$ fz^n
$-|-$gz$^{2n}-|-$hz^{3n} &c. & inde per Prop. 1. fit $\dot{R} =$
$_n \dot{f} zz^{n-1} -|- _{2n} g\dot{z} z^{2n-1} + _{3n} h\dot{z}z^{3n-1} -|-$ &c. quibus ſubſtitu-
tis & ſcripta BE ſeu 1 pro $\dot{z}$, fiet
$\theta e \overset{+\theta}{_{-|-\lambda n}} -|-\overset{-\theta}{f}z^n \overset{+\theta}{_{+2\lambda n}} gz^{2n} \overset{-\theta}{_{-|-3\lambda n}} hz^{3n} -|-$ &c. in $z^{\theta-1} R^{\lambda-1} = v = BC.$
Q. E. D.

PROP.

PROP. IV. THEOR. II.

Si Curvæ abſciſſa A B ſit z, & ſi pro $e + fz^n + gz^{2n}$ + &c. ſcribatur R, & pro $k + lz^n + mz^{2n} +$ &c. ſcribatur S ; ſit autem area Curvæ $z^\theta R^\lambda S^\mu$: erit ordinatim applicata BC =

$$\left.\begin{array}{l} \theta ek \genfrac{}{}{0pt}{}{+\theta}{+\lambda n} fk\ z^n \genfrac{}{}{0pt}{}{+\theta}{+2\lambda n} gkz^{2n} \cdot * \cdots \cdots * \cdots \cdots \\[6pt] \genfrac{}{}{0pt}{}{+\theta}{-\mu n} el\ z^n \genfrac{}{}{0pt}{}{+\theta}{\substack{-\lambda n \\ -\mu n}} fl z^{2n} \genfrac{}{}{0pt}{}{+\theta}{\substack{-2\lambda n \\ +\mu n}} gl z^{3n} * \cdots \cdots \\[6pt] \qquad \genfrac{}{}{0pt}{}{+\theta}{+2\mu n} emz^{2n} \genfrac{}{}{0pt}{}{+\theta}{\substack{-\lambda n \\ -2\mu n}} fmz^{3n} \genfrac{}{}{0pt}{}{+\theta}{\substack{+2\lambda n \\ +2\mu n}} gmz^{4n} \end{array}\right\} in\ z^{\theta-1}R^{\lambda-1}S^{\mu-1}$$

Demonſtratur ad modum Propoſitionis ſuperioris.

PROP. V. THEOR. III.

Si Curvæ abſciſſa A B ſit z, & pro $e + fz^n + gz^{2n} + hz^{3n} +$ &c. ſcribatur R : ſit autem ordinatim applicata $z^{\theta-1}R^{\lambda-1}$ in $a + bz^n + cz^{2n} + dz^{3n} +$ &c. & ponatur $\frac{\theta}{n} = r$. $r + \lambda = s$. $s + \lambda = t$. $t + \lambda = v$. &c. erit area

$$z^\theta R^\lambda\ in\ \frac{\frac{1}{n}a}{re} + \frac{\frac{1}{n}b - sfA}{\overline{r+1},e}z^n + \frac{\frac{1}{n}c \genfrac{}{}{0pt}{}{-s}{-1}fB - tgA}{\overline{r+2},e}z^{2n} + \frac{\frac{1}{n}d \genfrac{}{}{0pt}{}{-s}{-2}fC \genfrac{}{}{0pt}{}{-t}{-1}gB - vhA}{\overline{r+3},e}z^{3n} +$$

$$+ \frac{\genfrac{}{}{0pt}{}{-s}{-3}fD \genfrac{}{}{0pt}{}{-t}{-2}gC \genfrac{}{}{0pt}{}{-v}{-1}hB}{\overline{r+4},e}z^{4n} +$$ &c. Ubi A, B, C, D, &c.

A a a   denotant

denotant totas coefficientes datas terminorum fingu-
lorum in ferie cum fignis fuis $+$ & $-$, nempe A primi
termini coefficientem $\frac{\frac{1}{n}a}{re}$, B fecundi coefficientem

$\frac{\frac{1}{n}b - sfA}{\overline{r+1},e}$, C tertii coefficientem $\frac{\frac{1}{n}c \frac{-s}{-1}fB - tgA}{\overline{r+2},e}$, &

fic deinceps.

Demonſtratio.

Sunto juxta Propofitionem tertiam,

Curvarum Ordinatæ	& earundem areæ.

1. $\theta eA \frac{+\theta}{+\lambda n}fAz^n \frac{+\theta}{-2\lambda n}gAz^{2n} \frac{+\theta}{-3\lambda n}hAz^{3n}$ &c. $\quad\quad Az^\theta R^\lambda$.

2. $\cdots \overline{\theta + n}, eBz^n \frac{+\overline{\theta + n}}{+\lambda n}fBz^{2n} \frac{+\theta + n}{-2\lambda n}gBz^{3n}$ &c. $\quad Bz^{\theta + n} R^\lambda$.

3. $\cdots\cdots \overline{+\theta + 2n}, eCz^{2n} \frac{+\theta + 2n}{+\lambda n}fCz^{3n}$ &c. $\quad\quad Cz^{\theta + 2n} R^\lambda$.

4. $\cdots\cdots\cdots \overline{+\theta + 3n}, eDz^{3n}$ &c. $\quad\quad\quad Dz^{\theta + 3n} R^\lambda$.

$$z^{\theta - 1} R^{\lambda - 1} .$$

Et fi fumma ordinatarum ponatur æqualis ordi-
natæ $a + bz^n + cz^{2n} + dz^{3n} +$ &c. in $z^{\theta - 1} R^{\lambda - 1}$, fumma
arearum $z^\theta R^\lambda$ in $A + Bz^n + Cz^{2n} + Dz^{3n} +$ &c. æqualis erit areæ Curvæ cujus iſta eſt ordinata. Æquen-
tur igitur Ordinatarum termini correſpondentes, &
fiet $a = \theta eA$, $b = \frac{\theta}{+\lambda n}fA \frac{+\theta}{n}eB$, $c = \frac{\theta}{+2\lambda n}gA \frac{+\theta + n}{+\lambda n}fB$

$+ \overline{\theta + 2n}, eC$ &c. & inde $\frac{a}{\theta e} = A$. $\frac{b - \overline{\theta + \lambda n}, fA}{\overline{\theta + n}, e} = B$.

$\frac{c - \overline{\theta + 2\lambda n}, gA - \overline{\theta + n + \lambda n}, fB}{\overline{\theta + 2n}, e} = C$. Et fic deinceps in infinitum.

nitum. Pone jam $\frac{\theta}{"}=$ r. r$+_\wedge=$ s. s$+_\wedge=$ t &c. & in area $z^\theta R^\wedge \times \overline{A+Bz^n+Cz^{2n}+Dz^{3n}}$ &c. fcribe ipforum A, B, C, &c. valores inventos & prodibit feries propofita. Q. E. D.

Et notandum eft quod Ordinata omnis duobus modis iu feriem refolvitur. Nam index $"$ vel affirmativus eft poteft vel negativus. Proponatur Ordinata $\frac{3\,k-1\,z\,z}{z\,z\,\sqrt{k\,z-1\,z\,3+m\,z\,4}}$. Hæc vel fic fcribi poteft

$z^{-\frac{1}{2}}\times \overline{3k-1zz}\times \overline{k-1zz+mz3}|^{-\frac{1}{2}}$, vel fic $z\times-\overline{1+3kz^{-2}}$ $\times \overline{m-1z^{-1}+kz^{-3}}-\frac{1}{2}$. In cafu priore eft a $=$ 3k. b$=$0. c$=$–1. e$=$k. f$=$0. g$=$ –1. h$=$m. $_\wedge=-\frac{1}{2}$. $"=$1. $\theta-$1$=-\frac{5}{2}$. $\theta=-\frac{3}{2}=$r. s$=$–1. t$=-\frac{1}{2}$. v$=$0. In pofteriore eft a $=$ –1. b$=$0. c$=$3k. e$=$m. f$=$–1. g$=$0. h$=$1. $_\wedge=-\frac{1}{2}$. $"=$–1. $\theta-$1$=$1. $\theta=$2. r$=$–2. s$=$–1$\frac{1}{2}$. t$=$–1. v$=-\frac{1}{2}$. Tentandus eft cafus uterque. Et fi ferierum alterutra ob terminos tandem deficientes abrumpitur ac terminatur, habebitur area Curvæ in terminis finitis. Sic in exempli hujus priore cafu fcribendo in ferie valores ipforum a, b, c, e, f, g, h, $_\wedge$, θ, r, s, t, v, termini omnes poft primum evanefcunt in infinitum & area Curvæ prodit $-2\sqrt{\frac{k-1zz+mz3}{z3}}$. Et hæc area ob fignum negativum adjacet abfciffæ ultra ordinatam productæ. Nam area omnis affirmativa adjacet tam abfciffæ quam ordinatæ, negativa vero cadit ad contrarias partes ordinatæ & adjacet abfciffæ productæ, manente fcilicet figno Ordinatæ. Hoc modo feries alterutra & nonnunquam utraque femper terminatur & finita evadit fi Curva geometrice quadrari poteft. At fi Curva talem quadraturam non admittit, feries utraq; continuabitur in infinitum, & ea-

rum

rum altera converget & aream dabit approximando, præterquam ubi r (propter aream infinitam) vel nihil eft vel numerus integer & negativus, vel ubi $\frac{z}{c}$. æqualis eft unitati. Si $\frac{z}{e}$ minor eft unitate, converget feries in qua index „ affirmativus eft: fin $\frac{z}{e}$ unita te major eft, converget feries altera. In uno cafu area adjacet abfciflæ ad ufq; ordinatam ductæ, in altero adjacet abfciflæ ultra ordinatam productæ.

Nota infuper quod fi Ordinata contentum eft fub factore rationali Q & factore furdo irreducibili R^{π}, & factoris furdi latus R non dividit factorem rationalem Q; erit $\lambda - 1 = \pi$ & $R^{\lambda-1} = R^{\pi}$. Sin factoris furdi latus R dividit factorem rationalem femel, erit $\lambda - 1 = \pi + 1$ & $R^{\lambda-1} = R^{\pi+1}$: fi dividit bis, erit $\lambda - 1 = \pi + 2$ & $R^{\lambda-1} = R^{\pi+2}$: fi ter, erit $\lambda - 1 = \pi + 3$, & $R^{\lambda-1} = R^{\pi+3}$: & fic deinceps.

Si Ordinata eft fractio rationalis irreducibilis cum Denominatore ex duobus vel pluribus terminis compofito : refolvendus eft denominator in divifores fuos omnes primos. Et fi divifor fit aliquis cui nullus alius eft æqualis, Curva quadrari nequit : Sin duo vel plures fint divifores æquales, rejiciendus eft eorum unus, & fi adhuc alii duo vel plures fint fibi mutuo æquales & prioribus inæquales, rejiciendus eft etiam eorum unus, & fic in aliis omnibus æqualibus fi adhuc plures fint : deinde divifor qui relinquitur vel contentum fub divifioribus omnibus qui relinquuntur, fi plures funt, ponendum eft pro R, & ejus quadrati reciprocum R^{-2} pro $R^{\lambda-1}$, præterquam ubi contentum illud eft quadratum vel cubus vel quadrato quadratum,&c. quo cafu ejus latus

ponen-

ponendum eft pro R & poteftatis index 2 vel 3 vel 4 negative fumptus pro λ. & Ordinata ad denomina‑torem R^2 vel R^3 vel R^4 vel R^5 &c. reducenda.

Ut fi ordinata fit $\frac{z^5+z^4-8z^3}{z^5+z^4-5z^3-z^2+8z-4}$; quoniam hæc fractio irreducibilis eft & denominatoris diviſores funt pares, nempe $z-1$, $z-1$, $z-1$ & $z+2$, $z+2$, rejicio magnitudinis utriuſque diviſorem unum & reliquorum $z-1$, $z-1$, $z+2$ conten‑tum z^3-3z+2 pono pro R & ejus quadrati re‑ciprocum $\frac{1}{R^2}$ feu R^{-2} pro $R^{\lambda-1}$. Dein Ordina‑tam ad denominatorem R^2 feu $R^{1-\lambda}$ reduco, & fit

$$\frac{z^6-9z^4+8z^3}{z^3-3z+2|\,\text{quad.}},$$ id eft $z^3 \times \overline{8-9z} + z^3 \times \overline{2-3z+z^3}|^2$

Et inde eft $a=8$. $b=-9$. $c=0$. $d=-1$, &c. $e=2$. $f=-3$. $g=0$. $h=1$. $\lambda-1=-2$. $\lambda=-1$. $\mu=1$. $\theta-1=3$. $\theta=4=r$. $s=3$. $t=2$. $v=1$. Et his in ſerie ſcriptis prodit area $\frac{z^4}{z^3-3z+2}$, terminis om‑nibus in tota ſerie poſt primum evaneſcentibus.

Si deniq; Ordinata eſt fractio irreducibilis & ejus denominator contentum eſt ſub factore rationali Q & factore ſurdo irreducibili R^{π}, inveniendi ſunt la‑teris R diviſores omnes primi, & rejiciendus eſt di‑viſor unus magnitudinis cujuſq; & per diviſores qui reſtant, ſiqui ſint, multiplicandus eſt factor rationalis Q : & ſi factum æquale eſt lateri R vel lateris illius poteſtati alicui cujus index eſt numerus integer, eſto index ille m, & erit $\lambda-1=-\pi-m$, & $R^{\lambda-1}=R^{-\pi-m}$. Ut ſi Ordinata ſit $\frac{3q^5-q+x+9q^3xx-qqx^3-6qx^4}{qq-xx\sqrt{\text{cub.}\,q^3+qqx-qxx-x^3}}$,

quoniam.

quoniam factoris furdi latus R feu $q_3 + qqx - qxx - x_3$
divifores habet $q + x$, $q + x$, $q - x$ qui duarum funt
magnitudinum, rejicio diviforem unum magnitudinis utriufq; & per diviforem $q + x$ qui relinquitur
multiplico factorem rationalem $qq - xx$. Et quoniam factum $q_3 + qqx - qxx - x_3$ æquale eft la-
teri R, pono $m = 1$. & inde, cum π fit $\frac{1}{3}$, fit $\lambda - 1 = -\frac{4}{3}$.
Ordinatam igitur reduco ad denominatorem $R^{-\frac{4}{3}}$

& fit $Z^o \times \overline{3q^6 + 2q^5x + 8q^4xx + 8q^3x^3 - 7qqx^4 - 6qx^5}$
$\times \overline{q^3 + qqx - qxx - x^3}|^{-\frac{4}{3}}$. Unde eft $a = 3q^6$. $b = 2q^5$ &c.
$e = q_3$. $f = qq$ &c. $\theta - 1 = 0$. $\theta = 1 = n$. $\lambda = -\frac{1}{3}$. $r = 1$.
$s = \frac{2}{3}$. $t = \frac{1}{3}$. $v = 0$. Et his in ferie fcriptis prodit

area $\dfrac{3qqx + 3x^3}{\sqrt{cub.\ a_3 + aax - axx - x^3}}$, terminis omnibus in ferie tota
poft tertium evanefcentibus.

PROP. VI. THEOR. IV.

Si Curvæ abfciffa AB fit z, & fcribantur R pro
$e + fz^n + gz^{2n} + hz^{3n} + $ &c. & S pro $k + lz^n + mz^{2n}$
$+ nz^{3n}$ &c. fit autem ordinatim applicata $z^{\theta-1} R^{\lambda-1} S^{\mu-1}$
in $a + bz^n + cz^{2n} + dz^{3n}$ &c. & fi terminorum, e, f,
g, h, &c. & k, l, m, n. &c. rectangula fint.

ek	fk	gk	hk &c.
el	fl	gl	hl &c.
em	fm	gm	hm &c.
en	fn	gn	hn &c.

Et

Et ſi rectangulorum illorum coefficientes nume-
rales ſint reſpective

$$\tfrac{1}{n}\theta = \mathrm{r}. \quad \mathrm{r}\text{-}|\text{-}\lambda = \mathrm{s}. \quad \mathrm{s}\text{-}|\text{-}\lambda = \mathrm{t}. \quad \mathrm{t}\text{-}|\text{-}\lambda = \mathrm{v}. \ \&\mathrm{c}.$$

$$\mathrm{r}\text{-}|\text{-}\mu = \acute{\mathrm{s}}. \quad \mathrm{s}\text{-}|\text{-}\mu = \acute{\mathrm{t}}. \quad \mathrm{t}\text{-}|\text{-}\mu = \acute{\mathrm{v}}. \quad \mathrm{v}\text{-}|\text{-}\mu = \acute{\mathrm{w}}. \&\mathrm{c}.$$

$$\acute{\mathrm{s}}\text{-}|\text{-}\mu = \acute{\mathrm{t}}''. \quad \acute{\mathrm{t}}\text{-}|\text{-}\mu = \acute{\mathrm{v}}''. \quad \acute{\mathrm{v}}\text{-}|\text{-}\mu = \acute{\mathrm{w}}''. \quad \acute{\mathrm{w}}\text{-}|\text{-}\mu = \acute{\mathrm{x}}''. \&\mathrm{c}.$$

$$\mathrm{t}''\text{-}|\text{-}\mu = \mathrm{v}'''. \quad \mathrm{v}''\text{-}|\text{-}\mu = \mathrm{w}'''. \quad \mathrm{w}''\text{-}|\text{-}\mu = \mathrm{x}'''. \quad \mathrm{x}''\text{-}|\text{-}\mu = \mathrm{y}'''. \&\mathrm{c}.$$

area Curvæ erit hæc

$$z^{\theta}R^{\lambda}S^{\mu} \ \mathrm{in} \ \ \frac{\tfrac{1}{n}a}{r\,c\,k} + \frac{\tfrac{1}{n}b \ {-s\,f\,k \atop -s'\,e\,l}A}{\overline{r\text{-}|\text{-}1},\,e\,k}z^{n} + \frac{\tfrac{1}{n}c \ {-s\,\overline{\text{-}|\text{-}1},\,f\,k \atop -s'\text{-}|\text{-}1,\,e\,l}}B \ {-t\,g\,k \atop -t'\,f\,l \atop -t''\,e\,m}A}{\overline{r\text{-}|\text{-}2},\,e\,k}z^{2n} +$$

$$\text{-}|\text{-} \ \frac{\tfrac{1}{n}d \ {-s\text{-}|\text{-}2,\,f\,k \atop -s'\text{-}|\text{-}2,\,e\,l}C \ {-t\,\overline{\text{-}|\text{-}1},\,g\,k \atop -t'\text{-}|\text{-}1,\,f\,l \atop -t''\text{-}|\text{-}1,\,e\,m}}B \ {-v\,h\,k \atop -v'\,g\,l \atop -v''\,f\,m \atop -v'''\,e\,n}A}{\overline{r\text{-}|\text{-}3},\,e\,k}z^{3n} \text{-}|\text{-} \ \&\mathrm{c}.$$

Ubi A denotat termini primi coefficientem datam
$\frac{\tfrac{1}{n}a}{r\,c\,k}$ cum ſigno ſuo -|- vel —, B coefficientem datam
ſecundi, C coefficientem datam tertii, & ſic deinceps.
Terminorum vero, a, b, c, &c. k, l, m, &c. unus
vel plures deeſſe poſſunt. Demonſtratur Propoſitio
ad modum præcedentis, & quæ ibi notantur hic ob-
tinent. Pergit autem ſeries talium Propoſitionum in
infinitum, & Progreſſio ſeriei manifeſta eſt.

PROP.

PROP. VII. THEOR. V.

Si pro $e + fz^n + gz^{2n} +$ &c. fcribatur R ut fupra, & in Curvæ alicujus Ordinata $z^{\theta \pm n\sigma} R^{\lambda \pm \tau}$ maneant quantitates datæ θ, n, λ, e, f, g, &c. & pro σ ac τ fcribantur fucceffive numeri quicunq; integri : & fi detur area unius ex Curvis quæ per Ordinatas innumeras fic prodeuntes defignantur fi Ordinatæ funt duorum nominum in vinculo radicis, vel fi dentur areæ duarum ex Curvis fi Ordinatæ funt trium nominum in vinculo radicis, vel areæ trium ex Curvis fi Ordinatæ funt quatuor nominum in vinculo radicis, & fic deinceps in infinitum : dico quod dabuntur areæ curvarum omnium. Pro nominibus hic habeo terminos omnes in vinculo radicis tam deficientes quam plenos quorum indices dignitatum funt in progreffione arithmetica. Sic ordinata $\sqrt{a^4 - ax^3 + x^4}$ ob terminos duos inter a^4 & $-ax^3$ deficientes pro quinquinomio haberi debet. At $\sqrt{a^4 + x^4}$ binomium eft & $\sqrt{a^4 + x^4 - \frac{x^8}{a^4}}$ trinonium, cum progreffio jam per majores differentias procedat. Propofitio vero fic demonftratur.

CAS. I.

Sunto Curvarum duarum Ordinatæ $pz^{\theta-1} R^{\lambda-1}$ & $qz^{\theta+n-1} R^{\lambda-1}$, & areæ pA & qB, exiftente R quantitate trium nominum $e + fz^n + gz^{2n}$. Et cum per
Prop.

Prop. III. fit $z^\theta R^\lambda$ area curvæ cujus Ordinata eft $\theta e \genfrac{}{}{0pt}{}{+\theta}{-\vert-\lambda n} f z^n \genfrac{}{}{0pt}{}{+\theta}{+2\lambda n} g z^{2n}$ in $z^{\theta-1} R^{\lambda-1}$, fubduc Ordinatas & areas priores de area & Ordinata pofteriori, & manebit $\theta e \genfrac{}{}{0pt}{}{+\theta}{+\lambda n} \genfrac{}{}{0pt}{}{}{-p} f z^n \genfrac{}{}{0pt}{}{+\theta}{+2\lambda n} g z^{2n}$ in $z^{\lambda-1} R^{\lambda-1}$ Ordinata nova Curvæ,&
$-q z^n$

$z^\theta R^\lambda - pA - qB$ ejufdem area. Pone $\theta e = p$ & $\theta f \vert \lambda n f = q$ & Ordinata evadet $\genfrac{}{}{0pt}{}{\theta}{+2\lambda n} g z^{2n}$ in $z^{\theta-1} R^{\lambda-1}$, & area $z^\theta R^\lambda - \theta e A - \theta f B - \lambda n f B$. Divide utramq; per $\theta g \vert 2 \lambda n g$, & aream prodeuntem dic C, & affumpta utcunq; r, erit $r C$ area Curvæ cujus Ordinata eft $r z^{\theta + 2n - 1} R^{\lambda-1}$. Et qua ratione ex areis pA & qB aream rC Ordinatæ $r z^{\theta + 2n - 1} R^{\lambda-1}$ congruentem invenimus, licebit ex areis qB & rC aream quartam puta sD, ordinatæ $s z^{\theta + 3n - 1} R^{\lambda-1}$ congruentem invenire, & fic deinceps in infinitum. Et par eft ratio progreffionis ab areis B & A in partem contrariam pergentis. Si terminorum $\theta, \theta \vert \lambda n$, & $\theta \vert 2 \lambda n$ aliquis deficit & feriem abrumpit, affumatur area pA in principio progreffionis unius & area qB in principio alterius, & ex his duabus areis dabuntur areæ omnes in progreffione utraque. Et contra, ex aliis duabus areis affumptis fit regreffus per analyfin ad areas A & B, adeo ut ex duabus datis cæteræ omnes dentur. Q. E. O. Hic eft cafus Curvarum ubi ipfius z index θ augetur vel diminuitur perpetua additione vel fubductione quantitatis n. Cafus alter eft Curvarum ubi index λ augetur vel diminuitur unitatibus.

CAS.

C A S. II.

Ordinatæ $pz^{\theta-1}R^\lambda$ & $qz^{\theta+\eta-1}R^\lambda$, quibus areæ pA & qB jam refpondeant, fi in R feu $e+fz^\eta+gz^{2\eta}$ ducantur ac deinde ad R viciffim applicentur, evadunt pe $+ pfz^\eta + pgz^{2\eta} \times z^{\theta-1}R^{\lambda-1}$ & $qez^\eta + qfz^{2\eta} + qgz^{3\eta} \times z^{\theta-1}R^{\lambda-1}$. Et per Prop. III. eft $az^\theta R^\lambda$ area Curvæ cujus Ordinata eft $\theta ae \frac{+\theta}{+\lambda\eta}afz^\eta \frac{+\theta}{-2\lambda\eta}agz^{2\eta}$ in $z^{\theta-1}R^{\lambda-1}$, & $bz^{\theta+\eta}R^\lambda$ area Curvæ cujus ordinata eft $\frac{\theta}{+\eta}bez^\eta \frac{+\theta}{+\lambda\eta}bfz^{2\eta} \frac{+\theta}{-2\lambda\eta}bgz^{3\eta}$ in $z^{\theta-1}R^{\lambda-1}$. Et harum quatuor arearum fumma eft $pA+qB+az^\theta R^\lambda+bz^{\theta+\eta}R^\lambda$ & fumma refpondentium ordinatarum

$$\begin{array}{lllll} \theta ae & \frac{+\theta}{+\lambda\eta}afz^\eta & \frac{+\theta}{-2\lambda\eta}agz^{2\eta} & \frac{+\theta}{-2\lambda\eta} bgz^{3\eta} & \text{in } z^{\theta-1}R^{\lambda-1} \\ +pe & \frac{+\theta}{+\eta}be & \frac{+\theta}{+\eta}bf & +qg & \\ & & \frac{}{-\lambda\eta} & & \\ & +pf & +pg & & \\ & +qe & +qf & & \end{array}$$

Si terminus primus tertius & quartus ponantur feorfim æquales nihilo, per primum fiet $\theta ae + pe = o$ feu $-\theta a = p$, per quartum $-\theta b - \eta b - 2\lambda\eta b = q$, & per tertium (eliminando p & q) $\frac{2ag}{f} = b$. Unde fecundus fit $\frac{\lambda\eta aff - 4\lambda\eta age}{f}$, adeoq; fumma quatuor Ordinatarum eft $\frac{\lambda\eta aff - 4\lambda\eta age}{f} z^{\theta+\eta-1}R^{\lambda-1}$, & fumma totidem refpondentium arearum eft $a z^\theta R^\lambda + \frac{2ag}{f}z^{\theta+\eta}R_\lambda - \theta aA - \frac{2\theta+2\eta+4\lambda\eta}{f}agB$.

Divi-

Dividantur hæ fummæ per $\frac{\lambda n a f f - 4 \lambda n a g e}{f}$, & fi Quotum pofterius dicatur D, erit D area curvæ cujus ordinata eft Quotum prius $z^{\theta + \pi - 1} R^{\lambda - 1}$. Et eadem ratione ponendo omnes Ordinatæ terminos præter primum æquales nihilo poteft area Curvæ inveniri cujus Ordinata eft $z^{\theta - 1} R^{\lambda - 1}$. Dicatur area ifta C, & qua ratione ex areis A & B inventæ funt areæ C ac D, ex his areis C ac D inveniri poffunt aliæ duæ E & F ordinatis $z^{\theta - 1} R^{\lambda - 2}$ & $z^{\theta + \pi - 1} R^{\lambda - 2}$ congruentes, & fic deinceps in infinitum. Et per analyfin contrariam regredi licet ab areis E & F ad areas C ac D, & inde ad areas A & B, aliafq; quæ in progreffione fequuntur. Igitur fi index λ perpetua unitatum additione vel fubductione augeatur vel minuatur, & ex areis quæ Ordinatis fic prodeuntibus refpondent duæ fimpliciffimæ habentur ; dantur aliæ omnes in infinitum. Q. E. O.

C A S. III.

Et per cafus hofce duos conjunctos, fi tam index θ perpetua additione vel fubductione ipfius π, quam index λ perpetua additione vel fubductione unitatis, utcunq; augeatur vel minuatur, dabuntur arcæ fingulis prodeuntibus Ordinatis refpondentes. Q. E. O.

C A S.

C A S. IV.

Et fimili augmento fi ordinata conftat ex quatuor nominibus in vinculo radicali & dantur tres
arearum, vel fi conftat ex quinq; nominibus &
dantur quatuor arearum, & fic deinceps : dabuntur areæ omnes quæ addendo vel fubducendo nume-
rum n indici θ vel unitatem indici λ generari poffunt.
Et par eft ratio Curvarum ubi ordinatæ ex binomiis
conflantur, & area una earum quæ non funt geometrice quadrabiles datur.   Q. E. O.

PROP. VIII. THEOR. VI.

Si pro $e + fz^n + gz^{2n} + \&c.$ & $k + lz^n + mz^{2n} + \&c.$
fcribantur R & S ut fupra, & in Curvæ alicujus Or-
dinata $z^{\theta + n\sigma} R^{\lambda + \tau} S^{\mu + \nu}$ maneant quantitates datæ θ,
n, λ, μ, e, f, g, k, l, m, &c. & pro σ, τ, & ν, fcribantur fucceffive numeri quicunq; integri : & fi
dentur areæ duarum ex curvis quæ per ordinatas
fic prodeuntes defignantur fi quantitates R & S funt
binomia, vel fi dentur areæ trium ex curvis fi R
& S conjunctim ex quinq; nominibus conftant, vel
areæ quatuor ex curvis fi R & S conjunctim ex fex
nominibus conftant, & fic deinceps in infinitum :
dico quod dabuntur areæ curvarum omnium.

Demonftratur ad modum Propofitionis fuperioris.

PROP.

PROP. IX. THEOR. VII.

Æquantur Curvarum areæ inter se quarum Ordinatæ sunt reciproce ut fluxiones Absciffarum.

Nam contenta sub Ordinatis & fluxionibus Ab-
fcissarum erunt æqualia, & fluxiones arearum sunt
ut hæc contenta.

COROL. I.

Si assumatur relatio quævis inter Abscissas duarum Curvarum, & inde per Prop. 1. quæratur
relatio fluxionum Absciffarum, & ponantur Ordinatæ reciproce proportionales fluxionibus, inveniri
possunt innumeræ Curvæ quarum areæ sibi mutuo
æquales erunt.

COROL. II.

Sic enim Curva omnis cujus hæc est Ordinata
$z^{\theta-1}$ in $\overline{e + fz^{\eta} + gz^{2\eta} + \&c.}|^{\lambda}$ assumendo quantitatem
quamvis pro ν & ponendo $\frac{\eta}{\nu} = s$ & $z^{s} = x$, migrat in
aliam sibi æqualem cujus ordinata est $\frac{\nu}{\eta} x^{\frac{\nu\theta - \eta}{\eta}}$ in
$\overline{e + fx^{\nu} + gx^{2\nu} + \&c.}|^{\lambda}$.

COROL. III.

Et Curva omnis cujus Ordinata eft $z^{\theta-1}$ in $\overline{a + bz^n + cz^{2n} + \&c.} \times \overline{e + fz^n + gz^{2n} \&c.}|^\lambda$, affumendo quantitatem quamvis pro ν & ponendo $\frac{n}{\nu} = s$ & $z^s = x$, migrat in aliam fibi æqualem cujus ordinata eft $\frac{\nu}{n} x^{\frac{\nu\theta-n}{n}}$ in $\overline{a + bx^\nu + cx^{2\nu} + \&c.} \cdot x \overline{e + fx^\nu + gx^{2\nu} + \&c.}|^\lambda$.

COROL. IV.

Et Curva omnis cujus Ordinata eft $z^{\theta-1}$ in $\overline{a + bz^n + cz^{2n} + \&c.} \times \overline{e + fz^n + gz^{2n} + \&c.}|^\lambda \times \overline{k + lz^n + mz^{2n} + \&c.}|^\mu$, affumendo quantitatem quamvis pro ν & ponendo $\frac{n}{\nu} = s$ & $z^s = x$, migrat in aliam fibi æqualem cujus ordinata eft $\frac{\nu}{n} x^{\frac{\nu\theta-n}{n}}$ in $\overline{a + bx^\nu + cx^{2\nu} + \&c.} \cdot x \overline{e + fx^\nu + gx^{2\nu} + \&c.}|^\lambda x \overline{k + lx^\nu + mx^{2\nu} + \&c.}|^\mu$.

COROL. V.

Et Curva omnis cujus Ordinata eft $z^{\theta-1}$ in $\overline{e + fz^n + gz^{2n} + \&c.}|^\lambda$ ponendo $\frac{1}{z} = x$ migrat in aliam fibi æqualem cujus ordinata eft $\frac{1}{x^{\theta+1}} \times \overline{e + fx^n + gx^{-2n} + \&c.}|^\lambda$ id eft $\frac{1}{x^{\theta+1+n\lambda}} \times \overline{f + ex^n}|^\lambda$ fi duo funt nomina in vinculo radicis vel $\frac{1}{x^{\theta+1+m\lambda}} \times \overline{g + fx^n + ex^{2n}}|^\lambda$ fi tria funt nomina ; & fic deinceps.

CO-

COROL. VI.

Et Curva omnis cujus Ordinata eſt $z^{\theta-1}$ in $\overline{e + fz^n + gz^{2n} + \&c.}|^\lambda \times \overline{k + lz^n + mz^{2n} + \&c.}|^\mu$ ponendo $\frac{1}{z} = x$ migrat in aliam ſibi æqualem cujus ordinata eſt $\frac{1}{x^{\theta + 1}} \times \overline{e + fx^{-n} + gx^{-2n} + \&c.}|^\lambda$ $\times \overline{k + lx^{-n} + mx^{-2n} + \&c.}|^\mu$ id eſt $\frac{1}{x^{\theta + 1 + n\lambda + n\mu}} \times \overline{f + ex^n}|^\lambda$ $\times \overline{l + kx^n}|^\mu$ ſi bina ſunt nomina in vinculis radicum, vel $\frac{1}{x^{\theta + 1 + 2n\lambda + n\mu}} \times \overline{g + fx^n + ex^{2n}}|^\lambda \times \overline{l + kx^n}|^\mu$ ſi tria ſunt nomina in vinculo radicis prioris ac duo in vinculo poſterioris: & ſic in aliis. Et nota quod areæ duæ æquales in noviſſimis hiſce duobus Corollariis jacent ad contrarias partes ordinatarum. Si area in alterutra curva adjacet abſciſſæ, area huic æqualis in altera curva adjacet abſciſſæ productæ.

COROL. VII.

Si relatio inter Curvæ alicujus Ordinatam y & Abſciſſam z definiatur per æquationem quamvis ſectam hujus formæ, y^α in $e + fy^n z^\delta + gy^{2n} z^{2\delta} + hy^{3n} z^{3\delta}$ $+ \&c. = z^\beta$ in $k + ly^n z^\delta + my^{2n} z^{2\delta} + \&c.$ hæc figura aſſumendo $s = \frac{\alpha - \delta}{n}$, $x = \frac{1}{s} z^s$ & $\lambda = \frac{n - \delta}{\alpha\delta + \beta n}$, migrat in aliam ſibi æqualem cujus Abſciſſa x, ex data.

Ordinata

Ordinata v, determinatur per æquationem non affectam $\frac{1}{s}v^{\epsilon\lambda} \times \overline{e + fv^n + gv^{2n} + hv^{3n} + \&c.}^{\lambda} \times \overline{k + lv^n}$

$\overline{+ mv^{2n} + \&c.}^{-\lambda} = x.$

COROL. VIII.

Si relatio inter Curvæ alicujus Ordinatam y & Abfciffam z definitur per æquationem quamvis affectam hujus formæ, y^{α} in $\overline{e + fy^n z^{\delta} + gy^{2n} z^{2\delta} + \&c.}$ $= z^{\beta}$ in $\overline{k + ly^n z^{\delta} + my^{2n} z^{2\delta} + \&c.} + z^{\gamma}$ in $\overline{p + qy^n z^{\delta}}$ $\overline{+ ry^{2n} z^{2\delta} + \&c.}$ hæc figura affumendo $s = \frac{\alpha - \delta}{n}, x = \frac{1}{s} z^s,$ $\mu = \frac{\alpha\delta + \beta n}{n - \delta}$ & $\nu = \frac{\alpha\delta + \gamma n}{n - \delta}$, migrat in aliam fibi æqualem cujus Abfciffa x ex data Ordinata v determinatur per æquationem minus affectam v^{α} in $\overline{e + fv^n + gv^{2n}}$ $+ \&c. = s^{\mu} x^{\mu}$ in $\overline{k + lv^n + mv^{2n} + \&c.} + s^{\nu} x^{\nu}$ in $\overline{p + qv^n + rv^{2n} + \&c.}$

COROL. IX.

Curva omnis cujus Ordinata eft $\pi z^{\theta-1}$ in $\overline{e + \frac{1}{n} fz^n + \frac{1}{2n} gz^{2n} + \&c.} \times \overline{e + fz^n + gz^{2n} \&c.}^{\lambda-1} \times$ $\overline{a + b|ez^v + fz^{v+n} + gz^{v+2n} + \&c.|^{\eta}|^{\omega}}$, fi fit $\theta = \eta\lambda$ & affumantur $x = \overline{ez^v + fz^{v+n} + gz^{v+2n} + \&c.}^{\eta}$, $\sigma = \frac{\tau}{\eta}$ $\& \vartheta = \frac{\lambda - \tau}{\eta}$, migrat in aliam fibi æqualem cujus ordinata eft $x^{\vartheta} \times \overline{a + bx^{\sigma}}^{\omega}$. Et nota quod ordinata prior

in

in hoc Corollario evadit fimplicior ponendo $\lambda = 1$, vel ponendo $\tau = 1$ & efficiendo ut radix dignitatis extrahi poffit cujus index eft ω, vel etiam ponendo $\omega = -1$ & $\lambda = 1 = \tau = \sigma = \pi$, ut alios cafus præteream.

COROL. X.

Pro $ez^{\nu} + fz^{\nu+\eta} + gz^{\nu+2\eta} + \&c.$ $\nu ez^{\nu-1} \stackrel{+\nu}{\stackrel{\cdots}{\eta}} fz^{\nu+\eta-1}$ $\stackrel{+\nu}{\stackrel{+2\eta}{\cdots}} gz^{\nu+2\eta-1} + \&c.$ $k + lz^{\eta} + mz^{2\eta} + \&c.$ & $\eta lz^{\eta-1}$ $+ 2\eta m z^{2\eta-1} + \&c.$ fcribantur R, r, S & s refpective, & Curva omnis cujus ordinata eft $\pi Sr + \rho Rs$ in $R^{\lambda-1} S^{\mu-1}$ $\times \overline{aS^{\nu} + bR^{\eta}}\,|$, fi fit $\frac{\mu - \nu\omega}{\lambda} = \frac{\nu}{\tau} = \frac{\rho}{\pi}$, $\frac{\tau}{\pi} = \sigma$, $\frac{\lambda - \pi}{\sigma} = \vartheta$, & $R^{\pi} S^{\rho} = x$, migrat in aliam fibi æqualem cujus ordinata eft $x^{\vartheta} \times \overline{a + bx^{\sigma}}\,|^{\omega}$. Et nota quod Ordinata prior evadit fimplicior, ponendo unitates pro τ, ν, & λ vel μ, & faciendo ut radix dignitatis extrahi poffit cujus index eft ω, vel ponendo $\omega = -1$ vel $\mu = 0$.

PROP. X. PROB. III.

Invenire figuras fimpliciffimas cum quibus Curva quævis geometrice compari poteft, cujus ordinatim applicata y per æquationem non affectam ex data abfciffa z determinatur.

C A S.

C A S. I.

Sit Ordinata $az^{\theta-1}$, & area erit $\frac{1}{\theta}az^{\theta}$, ut ex Prop. V. ponendo $b = o = c = d = f = g = h$ & $e = 1$, facile colligitur.

C A S. II.

Sit Ordinata $az^{\theta-1} \times \overline{e + fz^{\eta} + gz^{2\eta}}\,\vert^{\lambda-1} + \&c.$ & si curva cum figuris rectilineis geometrice comparari poteft, quadrabitur per Prop. V. ponendo $b = o = c = d$. Sin minus convertetur in aliam curvam sibi æqualem cujus Ordinata eft $\frac{a}{\eta}x^{\frac{\theta-\eta}{\eta}} \times \overline{e + fx + gx^2 \&c.}\,\vert^{\lambda-1}$ per Corol. 2. Prop. IX. Deinde si de dignitatum indicibus $\frac{\theta-\eta}{\eta}$ & $\lambda-1$ per Prop. VII. rejiciantur unitates donec dignitates illæ fiant quam minimæ, devenietur ad figuras fimpliciffimas quæ hac ratione colligi poffunt. Dein harum unaquæq; per Corol. 5. Prop. IX. dat aliam quæ nonnunquam fimplicior eft. Et ex his per Prop. III. & Corol. 9 & 10, Prop. IX. inter fe collatis, figuræ adhuc fimpliciores quandoq; prodeunt. Deniq; ex figuris fimpliffimis affumptis facto regreffu computabitur area quæfita.

C A S. III.

Sit Ordinata $z^{\theta-1} \times \overline{a + bz^n + cz^{2n} + \&c.}$
$\times \overline{e + fz^n + gz^{2n} + \&c.}|^{\lambda-1}$, & hæc figura si quadrari
potest, quadrabitur per Prop. V. Sin minus, di-
stinguenda est ordinata in partes $z^{\theta-1} \times a \times \overline{e + fz^n}$
$\overline{+ gz^{2n} + \&c.}|^{\lambda-1}$, $z^{\theta-1} \times bz^n \times \overline{e + fz^n + gz^{2n} + \&c.}|^{\lambda-1}$,
&c. & per Cas. 2. inveniendæ sunt figuræ simpli-
cissimæ cum quibus figuræ partibus illis respon-
dentes comparari possunt. Nam areæ figurarum
partibus illis respondentium sub signis suis + & —
conjunctæ component aream totam quæsitam.

C A S. IV.

Sit Ordinata $z^{\theta-1} \times \overline{a + bz^n + cz^{2n} + \&c.} \times$
$\overline{e + fz^n + gz^{2n} + \&c.}|^{\lambda-1} \times \overline{k + lz^n + mz^{2n} + \&c.}|^{\mu-1}$:
& si Curva quadrari potest, quadrabitur per Prop. VI.
Sin minus, convertetur in simpliciorem per Corol. 4.
Prop. IX. ac deinde comparabitur cum figuris sim-
plicissimis per Prop. VIII. & Corol. 6, 9 & 10.
Prop. IX. ut fit in Casu 2 & 3.

C A S. V.

Si Ordinata ex variis partibus constat, partes
singulæ pro ordinatis curvarum totidem habendæ
sunt, & curvæ illæ quotquot quadrari possunt, sigilla-

tim

tim quadrandæ funt, earumq; ordinatæ de ordinata tota demendæ. Dein Curva quam ordinatæ pars refidua defignat feorfim (ut in Cafu 2, 3 & 4,) cum figuris fimpliciffimis comparanda eft cum quibus comparari poteft. Et fumma arearum omnium pro area Curvæ propofitæ habenda eft.

COROL. I.

Hinc etiam Curva .omnis cujus Ordinata eft radix quadratica affecta æquationis fuæ, cum figuris fimpliciffimis feu rectilineis feu curvilineis compari poteft. Nam radix illa ex duabus partibus femper conftat quæ feorfim fpectatæ non funt æquanum radices affectæ. Proponatur æquatio $aayy + zzyy = 2a^3y + {}^2z^3y - z^4$, & extracta radix erit $y = \dfrac{a^3 + z^{3+} a\sqrt{a^4 + {}^2az^3 - z^4}}{aa + zz}$ cujus pars rationalis $\dfrac{a3 + z3}{aa + zz}$ & pars irrationalis $\dfrac{a\sqrt{a^4 + 2az^3 - z^9}}{aa + zz}$ funt ordinatæ curvarum quæ per hanc Propofitionem vel quadrari poffunt vel cum figuris fimpliciffimis comparari cum quibus collationem geometricam admittunt.

COROL. II.

Et curva omnis cujus Ordinata per æquationem quamvis affectam definitur quæ per Corol. 7. Prop. IX. in æquationem non affectam migrat, vel quadratur

dratur per hanc Propofitionem fi quadrari poteft vel comparatur cum figuris fimpliciffimis cum quibus compari poteft. Et hac ratione Curva omnis quadratur cujus æquatio eft trium terminorum. Nam æquatio illa fi affecta fit tranfmutatur in non affectam per Corol. 7. Prop. IX. ac deinde per Corol. 2 & 5. Prop. IX. in fimplicffimam migrando, dat vel quadraturam figuræ fi quadrari poteft, vel curvam fimpliciffimam quacum comparatur.

COROL. III.

Et Curva omnis cujus Ordinata per æquationem quamvis affectam definitur quæ per Corol. 8. Prop. IX. in æquationem quadraticam affectam migrat; vel quadratur per hanc Propofitionem & hujus Corol. 1. fi quadrari poteft, vel comparatur cum figuris fimplieiffimis cum quibus collationem geometricam admittit.

SCHOLIUM.

Ubi quadrandæ funt figuræ; ad Regulas hafce generales femper recurrere nimis moleftum effet: præftat Figuras quæ fimpliciores funt & magis ufui effe poffunt femel quadrare & quadraturas in Tabulam referre, deinde Tabulam confulere quoties ejufmodi Curvam aliquam quadrare oportet. Hujus autem generis funt Tabulæ duæ fequentes, in quibus z denotat Abfciffam, y Ordinatam rectangulam

gulam & t Aream Curvæ quadrandæ, & d, e, f, g,
g. h, " funt quantitates datæ cum fignis fuis $+$ & $-$.

TABULA

Curvarum fimpliciorum quæ quadrari poffunt.

Curvarum formæ.   Curvarum areæ.

Forma prima.

$$dz^{n-1} = y. \qquad \tfrac{d}{n}z^n = t.$$

Forma fecunda.

$$\frac{dz^{n-1}}{ee+2efz_n+ffz^2} = y. \qquad \frac{dz_n}{nee+nefz_n} = t, \ vel \ \frac{-d}{nef+nffz_n} = t.$$

Forma tertia.

1. $dz_{-1}^{n} \sqrt{e+fz^n} = y.$ $\quad \frac{2d}{3nf} R^3 = t,$ exiftente $R = \sqrt{e+fz^n}$

2. $dz_{-1}^{2n} \sqrt{e+fz^n} = y.$ $\quad \frac{-4e+6fz_n}{15nff} dR^3 = t.$

3. $dz_{-1}^{3n} \sqrt{e+fz^n} = y.$ $\quad \frac{16ee-24efz_n+30ffz2_n}{105nf3} dR^3 = t.$

4. $dz_{-1}^{4n} \sqrt{e+fz^n} = y.$ $\quad \frac{-96e3+144eefz_n-180effz2_n+210f3z3_n}{945nf4} dR^3 = t.$

Forma quarta.

1. $\dfrac{dz^{n-1}}{\sqrt{e+fz_n}} = y.$ $\quad \frac{2d}{nf} R = t.$

2. $\dfrac{dz^{2n-1}}{\sqrt{e+fz_n}} = y.$ $\quad \frac{-4e+2fz_n}{3nff} d R = t.$

$$dz^{3n-1}$$

$$3.\ \frac{dz^{3n-1}}{\sqrt{e+fz_n}} = y.\qquad \frac{16ee-8efz_n+6ffz2_n}{15nf3}dR=t.$$

$$4.\ \frac{dz^{4n-1}}{\sqrt{e+fz_n}} = y.\qquad \frac{-96e3+48eefz_n-36effz2_n+30f323n}{105nf4}dR=t.$$

TABULA

Curvarum fimpliciorum quæ cum Ellipfi &
Hyperbola compari poffunt.

Sit jam a G D vel P G D vel G D S Sectio Conica cujus area ad Quadraturam Curvæ pro- *Fig.* 5,6,7,8. pofitæ requiritur, fitq; ejus centrum A, Axis K a, Vertex a, Semiaxis conjugatus A P, datum Abfciffæ principium A vel a vel α, Abfciffa A B vel a B vel α B = x, Ordinata rectangula B D = v, & Area A B D P vel a B D G vel α B D G = s, exiftente α G Ordinata ad punctum α. Jungantur K D, A D, a D. Ducatur Tangens D T occurrens Abfciffæ A B in T, & compleatur parallelogrammum A B D O. Et fiquando ad quadraturam Curvæ propofitæ requiruntur areæ duarum Sectionem Conicarum, dicatur pofterioris Abfciffa ξ, Ordinata Υ, & Area σ. Sit autem ÷ differentia duarum quantitatum ubi incertum eft utrum pofterior de priori an prior de pofteriori fubduci debeat.

Curva-

Curvarum Formæ. Sectionis Conicæ. Curvarum Areæ.

Forma prima.   Abscissa. Ordinata.

Fig. 5.

1. $\dfrac{dz^{n-1}}{e+fz^n}=y.$ $z^n=x.$ $\dfrac{d}{e+fx}=v.$ $\dfrac{1}{n}s=t=\dfrac{aGDB}{n}$

2. $\dfrac{dz^{2n-1}}{e+fz^n}=y.$ $z^n=x.$ $\dfrac{d}{e+fx}=v.$ $\dfrac{d}{nf}z^n-\dfrac{e}{nf}s=t.$

3. $\dfrac{dz^{3n-1}}{e+fz^n}=y.$ $z^n=x.$ $\dfrac{d}{e+fx}=v.$ $\dfrac{d}{2nf}z^{2n}-\dfrac{de}{nff}z^n+\dfrac{ee}{nff}s=t.$

Forma secunda.

Fig. 6,7.

1. $\dfrac{dz^{\frac{1}{2}n-1}}{e+fz^n}=y.$ $\sqrt{\dfrac{d}{e+fz_n}}=x.$ $\sqrt{\dfrac{d}{f}-\dfrac{e}{f}xx}=v.$ $\dfrac{2xv\div 4s}{n}=t=\dfrac{4}{n}ADGa.$

2. $\dfrac{dz^{\frac{3}{2}n-1}}{e+fz^n}=y.$ $\sqrt{\dfrac{d}{e+fz_n}}=x.$ $\sqrt{\dfrac{d}{f}-\dfrac{e}{f}xx}=v.$ $\dfrac{2de}{nf}z^{\frac{n}{2}}+\dfrac{4es-nexv}{nf}=t.$

3. $\dfrac{dz^{\frac{5}{2}n-1}}{e+fz^n}=y.$ $\sqrt{\dfrac{d}{e+fz_n}}=x.$ $\sqrt{\dfrac{d}{f}-\dfrac{e}{f}xx}=v.$ $\dfrac{2de}{3nf}z^{\frac{3n}{2}}-\dfrac{2dee}{nff}z^{\frac{n}{2}}+\dfrac{2eexv-4ees}{nff}=t.$

Forma

Forma tertia.

*Fig.*6,7,8. 1. $\frac{d}{2}\sqrt{e+fz_n}=y.\ \frac{1}{z_n}=xx.\ \sqrt{f+exx}=v.\ \frac{4de}{n^2}$ in $\frac{v_3}{2ex}-s=t=\frac{4de}{n^2}$ in aGDT, vel in APDB÷TDB.

Vel fic, $\frac{1}{2n}=x.\ \sqrt{fx+exx}=v.\ \frac{8dee}{n^2f}$ in $s-\frac{1}{2}xv-\frac{fv}{4e}+\frac{ffv}{4eex}=t=\frac{8dee}{n^2f}$ in aGDA$+\frac{ffv}{4eex}$.

2. $\frac{d}{z^{\frac{n}{+1}+1}}\sqrt{e+fz_n}=y.\ \frac{1}{z_n}=xx.\ \sqrt{f+exx}=v.\frac{-2d}{n}s=t=\frac{2d}{n}$APDB, feu $\frac{2d}{n}$aGDB.

Vel fic, $\frac{1}{z_n}=x.\ \sqrt{fx+exx}=v.\ \frac{4de}{n^2}$ in $s-\frac{1}{2}xv-\frac{fv}{2e}=t=\frac{4de}{n^2}\times$aGDK.

3. $\frac{d}{z^{\frac{2n}{+1}}}\sqrt{e+fz_n}=y.\frac{1}{2n}=x.\ \sqrt{fx+exx}=v.\ \frac{-d}{n}s=t=\frac{d}{n}\times$—aGDB vel BDPK.

4. $\frac{d}{z^{\frac{3n}{+1}}}\sqrt{e+fz_n}=y.\frac{1}{7n}=x.\ \sqrt{fx+exx}=v.\ \frac{3dfs-2dv_3}{6ne}=t.$

[201]

Forma quarta.

Fig. 6. 1. $\frac{d}{z\sqrt{e+fz_n}}=y.\frac{1}{z_n}=xx.\ \sqrt{f+exx}=v.\ \frac{4d}{n^2}$ in $\frac{1}{2}xv\div s=t=\frac{4d}{n^2}$ in PAD vel in aGDA.

Vel fic, $\frac{1}{z_n}=x.\ \sqrt{fx+exx}=v.\ \frac{8de}{n^2f}$ in $s-\frac{1}{2}xv-\frac{fv}{4e}=t=\frac{8de}{n^2f}$ in aGDA.

2. $\frac{d}{z^{\frac{n}{+1}+1}\sqrt{e+fz_n}}=y.\ \frac{1}{z_n}=xx.\ \sqrt{f+exx}=v.\ \frac{2d}{nc}$ in $s-xv=t=\frac{2d}{ne}$ in POD, vel in AODGa.

Vel fic, $\frac{1}{z_n}=x.\ \sqrt{fx+exx}=v.\ \frac{4d}{n^2}$ in $\frac{1}{2}xv\div s=t=\frac{4d}{n^2}$ in aDGa.

3. $2^{\frac{2n}{+1}}$

$3. \; \dfrac{d}{z_{+1}^{2n}\sqrt{e+fz_n}} = y. \; \dfrac{1}{2n} = x. \; \sqrt{fx+exx} = v. \; \dfrac{d}{nc} \text{ in } 3\,s \div 2xv = t = \dfrac{d}{nc} \text{ in } 3aDGa \div \triangle aDB.$

$4. \; \dfrac{d}{z_{+1}^{3n}\sqrt{e+fz_n}} = y. \; \dfrac{1}{2n} = x. \; \sqrt{fx+exx} = v. \; \dfrac{10dfxv - 15dfs - 2dexxv}{6ncc} = t.$

Forma quinta.

$1. \; \dfrac{dz_{n-1}}{e+fz_n+gz_{2n}} = y. \; \sqrt{\dfrac{d}{e+fz_n+gz_{2n}}} = x. \; \sqrt{\dfrac{d}{g} + \dfrac{ff-4eg}{4gg}xx} = v. \; \dfrac{xv-2s}{n} = t.$

Vel fic, $\sqrt{\dfrac{d2z_n}{e+fz_n+gz_{2n}}} = x. \; \sqrt{\dfrac{d}{e} + \dfrac{ff-4eg}{4ee}xx} = v. \; \dfrac{2s-xv}{n} = t.$

$2. \; \dfrac{d2z_{n-1}}{e+fz_n+gz_{2n}} = y. \begin{cases} \sqrt{\dfrac{d}{e+fz_n+gz_{2n}}} = x. \\ fz_n+gz_{2n} = \xi. \end{cases} \begin{cases} \sqrt{\dfrac{d}{g} + \dfrac{ff-4eg}{4gg}xx} = v. \\ \dfrac{1}{e+\xi} = \tau. \end{cases} \begin{cases} \dfrac{dg-1 fs-fxv}{2sg} = t. \end{cases}$

Forma fexta, ubi fcribitur p pro $\sqrt{ff-4eg}$.

$1. \; \dfrac{dz_{\frac{1}{2}n-1}}{e+fz_n+gz_{2n}} = y. \begin{cases} \sqrt{\dfrac{2dg}{f-p+2gz_n}} = x. \\ \sqrt{\dfrac{2dg}{f+p+2gz_n}} = \xi. \end{cases} \begin{cases} \sqrt{d + \dfrac{-f+p}{2g}xx} = v. \\ \sqrt{d + \dfrac{-f-p}{2g}\xi\xi} = \tau. \end{cases} \begin{cases} \dfrac{2xv-4s-2\xi\tau+4\sigma}{np} = t. \end{cases}$

$2. \; \dfrac{dz_{\frac{1}{2}n-1}}{e+fz_n+gz_{2n}} = y. \begin{cases} \sqrt{\dfrac{2dez_n}{f z_n - pz_n + 2e}} = x. \\ \sqrt{\dfrac{2dez_n}{f z_n + pz_n + 2e}} = \xi. \end{cases} \begin{cases} \sqrt{d + \dfrac{-f+p}{2e}xx} = v. \\ \sqrt{d + \dfrac{-f-p}{2e}\xi\xi} = \tau. \end{cases} \begin{cases} \dfrac{4s-2xv-4\sigma+2\xi\tau}{np} = t. \end{cases}$

Forma

Forma feptima.

$$1.\ \frac{d}{z}\sqrt{e+fz^n+gz^{2n}}=y.\ \begin{cases}z_n=x.\sqrt{e+fx+gxx}=v.\\ \frac{1}{z_n}=\xi.\sqrt{g+f\xi+e\xi\xi}=\tau.\end{cases}\ \frac{4deg\tau+2def\tau-2dffv-8dee\sigma+4dfgs}{4neg-nff}=t.$$

Fig. 6,7. $2.\ dz_n\sqrt{e+fz^n+gz^{2n}}=y.\ z^n=x.\ \sqrt{e+fx+gxx}=v.\ \frac{d}{n}s=t=\frac{d}{n}\ in\ \alpha GDB.$

$3.\ dz_{-1}^{2n}\sqrt{e+fz^n+gz^{2n}}=y.\ z^n=x.\ \sqrt{e+fx+gxx}=v.\ \frac{d}{3ng}v3-\frac{df}{2ng}s=t.$

$4.\ dz_{-1}^{3n}\sqrt{e+fz^n+gz^{2n}}=y.\ z^n=x.\ \sqrt{e+fx+gxx}=v.\ \frac{6dgx-5df}{24ngg}v3+\frac{5dff-4deg}{16ngg}s=t.$

Forma octava.

Fig. 6 $1.\ \frac{dz^{n-1}}{\sqrt{e+fz^n+gz^{2n}}}=y.z^n=x.\sqrt{e+fx+gxx}=v.\frac{8dgs-4dgxv-2dfv}{4neg-nff}=t=\frac{8dg}{4neg-nff}\ in\ \alpha GDB+\Delta DBA.$

$2.\ \frac{dz^{2n-1}}{\sqrt{e+fz^n+gz^{2n}}}=y.\ z^n=x.\ \sqrt{e+fx+gxx}=v.\ \frac{-4dfs+2dfxv+4dev}{4neg-nff}=t.$

$3.\ \frac{dz^{3n-1}}{\sqrt{e+fz^n+gz^{2n}}}=y.\ z^n=x.\ \sqrt{e+fx+gxx}=v.\ \frac{\frac{3dff}{-4deg}s\frac{-2dff}{+4deg}xv-2defv}{4neg-nffg}=t.$

$4.\ \frac{dz^{4n-1}}{\sqrt{e+fz^n+gz^{2n}}}=y.\ z^n=x.\ \sqrt{e+fx+gxx}=v.\ \frac{\frac{36defg}{-15deff}s\frac{+8degg}{-2dffg}xxv\frac{-28defg}{+10deff}xv\frac{+10deeg}{-16deet}v}{24neg3-6nffg}=t.$

Ddd 2

Forma

Forma nona.

1. $\dfrac{dz^n_{.1}\sqrt{e+fz^n}}{g+hz^n} = y.\ \sqrt{\dfrac{d}{g+hz_n}} = x.\ \sqrt{\dfrac{df}{h}+\dfrac{eh-fg}{h}xx} = v.\ \dfrac{\frac{4fg}{-4ch}S\frac{-2fg}{+2ch}xv+\frac{2dfv}{x}}{{}_nfh} = t.$

2. $\dfrac{dz^{2n}_{.1}\sqrt{e+fz^n}}{g+hz^n} = y.\sqrt{\dfrac{d}{g+hz_n}} = x.\sqrt{\dfrac{df}{h}+\dfrac{eh-fg}{h}xx} = v.\ \dfrac{\frac{4egh}{-4fgg}S\frac{-2egh}{+2fgg}xv+\frac{2}{3}dh\frac{v^3}{x^3}-2dfg\frac{v}{x}}{{}_nfhh} = t.$

Forma decima.

Fig. 6,7.   1. $\dfrac{dz^{n\cdot 1}}{g+hz_n\ \sqrt{e+fz_n}} = y.\ \sqrt{\dfrac{d}{g+hz_n}} = x.\ \sqrt{\dfrac{df}{h}+\dfrac{eh-fg}{h}xx} = v.\ \dfrac{2xv-4s}{gf} = t = \dfrac{4}{nf}ADGa.$

2. $\dfrac{dz^{2n-1}}{g+hz_n\ \sqrt{e+fz_n}} = y.\ \sqrt{\dfrac{d}{g+hz_n}} = x.\ \sqrt{\dfrac{df}{h}+\dfrac{eh-fg}{h}xx} = v.\ \dfrac{4gs-2gxv+\frac{2dv}{x}}{nfh} = t.$

Forma undecima.

1. $dz^{-1}\sqrt{\dfrac{e+fz_n}{g+hz_n}} = y. \begin{cases}\sqrt{g+hz^n} = x.\ \sqrt{\dfrac{eh-fg}{h}+\dfrac{f}{h}xx} = v.\\ \sqrt{h+gz^{-n}} = \xi.\ \sqrt{\dfrac{fg-eh}{g}+\dfrac{e}{g\xi\xi}} = \gamma.\end{cases} \begin{cases}\dfrac{dxv^3z^{-n}-4dfs-4des}{nfg-nch} = t.\end{cases}$

2. $dz^n_{.1}\sqrt{\dfrac{e+fz_n}{g+gz_n}} = y.\ \sqrt{g+hz^n} = x.\ \sqrt{\dfrac{eh-fg}{h}+\dfrac{f}{h}xx} = v.\ \dfrac{2d}{nh}s = t.$

3. $dz^{2n}_{.1}\sqrt{\dfrac{e+fz_n}{g+hz_n}} = y\ \sqrt{g+hz^n} = x.\ \sqrt{\dfrac{eh-fg}{h}+\dfrac{f}{h}xx} = v.\ \dfrac{dhxv^3\frac{-3dfg}{-deh}s}{2nfhh\gamma} = t.$

In

In Tabulis hifce, feries Curvarum cujufq; formæ utrinq; in infinitum continuari poteft. Scilicet in Tabula prima, in numeratoribus arearum formæ tertiæ & quartæ, numeri coefficientes initialium terminorum (2,—4, 16,—96, 868, &c.) generantur multiplicando numeros—2,—4,—6,—10,&c. in fe continuo, & fubfequentium terminorum coefficientes ex initialibus derivantur multiplicando ipfos gradatim, in Forma quidem tertia, per —$\frac{3}{2}$, —$\frac{5}{4}$, —$\frac{7}{6}$, —$\frac{9}{8}$, —$\frac{11}{10}$ &c. in quarta vero per —$\frac{1}{2}$, —$\frac{3}{4}$, —$\frac{5}{6}$, —$\frac{7}{8}$, —$\frac{9}{10}$, &c. Et Denominatorum coefficientes 3, 15, 105, &c. prodeunt multiplicando numeros 1, 3, 5, 7, 9, &c. in fe continuo.

In fecunda vero Tabula, feries Curvarum formæ primæ, fecundæ, quintæ, fextæ, nonæ & decimæ ope folius divifionis, & formæ reliquæ ope Propofitionis tertiæ & quartæ, utrinq; producuntur in infinitum.

Quinetiam hæ feries mutando fignum numeri „ variari folent. Sic enim, e. g. Curva $\frac{d}{2}\sqrt{e+fz^n}$ = y, evadit $\frac{d}{2\frac{1}{2}n-1}$ $\sqrt{f+ez^n}$.

PROP. IX. THEOR. VIII.

Sit ADIC Curva quævis Abfciffam habens *Fig. 9.* AB=z & Ordinatam BD=y, & fit AEKC Curva alia cujus Ordinata BE æqualis eft prioris areæ
ABC

ADB ad unitatem applicatæ, & AFLC Curva
tertia cujus Ordinata BF æqualis eft fecundæ areæ
AEB ad unitatem applicatæ, & AGMC Curva
quarta cujus Ordinata BG æqualis eft tertiæ areæ
AFB ad unitatem applicatæ, & AHNC Curva
quinta cujus Ordinata BH æqualis eft quartæ areæ
AGB ad unitatem applicatæ, & fic deinceps in
infinitum.   Et funto A, B, C, D, E, &c. Areæ Curvarum Ordinatas habentium y, zy, z^2y, z^3y, z^4y,
& Abfciffam communem z.

Detur Abfciffa quævis $AC = t$, fitq; $BC = t - z$
$= x$, & funto P, Q, R, S, T areæ Curvarum Ordinatas habentium x, xy, xxy, x^3y, x^4y & Abfciffam
communem x.

Terminenter autem hæ areæ omnes ad Abfciffam
totam datam AC, nec non ad Ordinatam pofitione
datam & infinite produ&am CI : & erit arearum
fub initio pofitarum prima $ADIC = A = P$, fecuna
$AEKC = tA - B = Q$. Tertia $AFLC = \frac{ttA - 2tB + C}{2} = \frac{1}{2}R$.
Quarta $AGMC = \frac{t^3A - 3ttB + 3tC - D}{6} = \frac{1}{6}s$.   Quinta
$AHNC = \frac{t^4A - 4t^3B + 6ttC - 4tD + E}{24} = \frac{1}{24} T$.

COROL.

Unde fi Curvæ quarum Ordinatæ funt y, zy, z³y, z³y, &c. vel y, xy, x²y, x³y, &c. quadrari poffunt, quadrabuntur etiam Curvæ ADIC, AEKC, AFLC, AGMC, &c. & habebuntur Ordinatæ BE, BF, BG, BH areis Curvarum proportionales.

SCHOLIUM.

Quantitatum fluentium fluxiones effe primas, fecundas, tertias, quartas, aliafq; diximus fupra. Hæ fluxiones funt ut termini ferierum infinitarum convergentium. Ut fi z^n fit quantitas fluens & fluendo evadat $\overline{z + o}^n$, deinde refolvatur in feriem convergentem $z^n + noz^{n-1} + \frac{nn-n}{2}ooz^{n-2} + \frac{n3 - 3nn + 2n}{6}o^3z^{n-3}$ + &c. terminus primus hujus feriei z^n erit quantitas illa fluens, fecundus noz^{n-1} erit ejus incrementum primum feu differentia prima cui nafcenti proportionalis eft ejus fluxio prima, tertius $\frac{nn-n}{2}oz^{n-2}$ erit ejus incrementum fecundum feu differentia fecunda cui nafcenti proportionalis eft ejus fluxio fecunda, quartus $\frac{n3 - 3nn + 2n}{6}o^3z^{n-3}$ erit ejus incrementum tertium feu differentia tertia cui nafcenti fluxio tertia proportionalis eft, & fic deinceps in infinitum.

Exponi

Exponi autem poffunt hæ fluxiones per Curvarum Ordinatas BD, BE, BF, BG, BH, &c. Ut fi Ordinata BE ($=\frac{ADB}{I}$) fit quantitas fluens, erit ejus fluxio prima ut ordinata BD. Si BF ($=\frac{AEB}{I}$) fit quantitas fluens, erit ejus fluxio prima ut Ordinata BE & fluxio fecunda ut Ordinata BD. Si BH ($=\frac{AGB}{I}$) fit quantitas fluens, erunt ejus fluxiones, prima, fecunda, tertia & quarta, ut Ordinatæ BG, BF, BE, BD refpective.

Et hinc in æquationibus quæ quantitates tantum duas incognitas involvunt, quarum una eft quantitas uniformiter fluens & altera eft fluxio quælibet quantitatis alterius fluentis, inveniri poteft fluens illa altera per quadraturam Curvarum. Exponatur enim fluxio ejus per Ordinatam BD, & fi hæc fit fluxio prima, quæratur area $ADB = BE \times I$, fi fluxio fecunda, quæratur area $AEB = BF \times I$, fi fluxio tertia, quæratur area $AFB = BG \times I$, &c. & area inventa erit exponens fluentis quæfitæ.

Sed & in æquationibus quæ fluentem & ejus fluxionem primam fine altera fluente, vel duas ejufdem fluentis fluxiones, primam & fecundam, vel fecundam & tertiam, vel tertiam & quartam, &c. fine alterutra fluente involvunt : inveniri poffunt fluentes per quadraturam Curvarum. Sit æquatio $a a \dot{v} = a v + v v$, exiftente $v = BE$, $\dot{v} = BD$, $z = AB$ & $\dot{z} = I$, & æquatio illa complendo dimenfiones fluxionum, evadet $a a \dot{v} = a v \dot{z}$ $+ v v \dot{z}$, feu $\frac{a a \dot{v}}{a v + v v} = \dot{z}$. Jam fluat v uniformiter &

fit.

fit ejus fluxio $\dot{v}=1$ & erit $\frac{aa}{a\dot{v}+\dot{v}\dot{v}}=\dot{z}$, & quadrando Curvam cujus Ordinata eft $\frac{aa}{a\dot{v}+\dot{v}\dot{v}}$ & Abfciffa v, habebitur fluens z. Adhæc fit æquatio $aa\dot{v}=a\dot{v}+\dot{v}\dot{v}$ exiftente $v=BF$, $\dot{v}=BE$, $\ddot{v}=BD$ & $z=AB$ & per relationem inter $\dot{v}$ & $\ddot{v}$ feu BD & BE invenietur relatio inter AB & BE ut in exemplo fuperiore. Deinde per hanc relationem invenietur relatio inter AB & BF quadrando Curvam AEB.

Æquationes quæ tres incognitas quantitates involvunt aliquando reduci poffunt ad æquationes quæ duas tantum involvunt, & in his cafibus fluentes invenientur ex fluxionibus ut fupra. Sit æquatio $a-bx^m=cxy^n\dot{y}+dy^{2n}\dot{y}\dot{y}$: Ponatur $y^n\dot{y}=\dot{v}$ & erit $a-bx^m cx\dot{v}+d\dot{v}\dot{v}$. Hæc æquatio quadrando Curvam cujus Abfciffa eft x & Ordinata $\dot{v}$ dat aream v, & æquatio altera $y^n\dot{y}=\dot{v}$ regrediendo ad fluentes dat $\frac{1}{n+1}y^{n+1}=v$. Unde habetur fluens y.

Quinetiam in æquationibus quæ tres incognitas involvunt & ad æquationes quæ duas tantum involvunt reduci non poffunt, fluentes quandoq; prodeunt per quadraturam Curvarum. Sit æquatio $\overline{ax^m+bx^n}|^p=rex^{r-1}y^s+sex^r\dot{y}y^{s-1}-f\dot{y}y^t$, exiftente $x=1$. Et pars pofterior $rex^{r-1}y^s+sex^r\dot{y}y^{s-1}-f\dot{y}y^t$, regrediendo ad fluentes, fit $ex^ry^s-\frac{f}{t+1}y^{t+1}$, quæ proinde eft ut area Curvæ cujus Abfciffa eft x & Ordinata $\overline{ax^m+bx^n}|^p$, & inde datur fluens y.

E e e · Sit

Sit æquatio $\dot{x} \times \overline{a\,x^m + bx^n}|^p = \frac{d\dot{y}\,y_{n-1}}{\sqrt{e + fy_n}}$. Et fluens cujus fluxio eft $\dot{x} \times \overline{a\,x^m + bx^n}|^p$ erit ut area Curvæ cujus Abfciffa eft x & Ordinata eft $\overline{a\,x^m + bx^n}|^p$. Item fluens cujus fluxio eft $\frac{d\dot{y}\,y_{n-1}}{\sqrt{e + fy_n}}$ erit ut area Curvæ cujus Abfciffa eft y & Ordinata $\frac{d\dot{y}\,y_{n-1}}{\sqrt{e + fy_n}}$, id eft (per Cafum 1. Formæ quartæ Tab. I.) ut area $\frac{2d}{nf}\overline{\sqrt{e + fy^n}}$. Pone ergo $\frac{2d}{nf}\sqrt{e + fy^n}$ æqualem areæ Curvæ cujus Abfciffa eft x & Ordinata $\overline{a\,x^m + bx^n}|^p$ & habebitur fluens y.

Et nota quod fluens omnis quæ ex fluxione prima colligitur augeri poteft vel minui quantitate quavis non fluente. Quæ ex fluxione fecunda colligitur augeri poteft vel minui quantitate quavis cujus fluxio fecunda nulla eft. Quæ ex fluxione tertia colligitur augeri poteft vel minui quantitate quavis cujus fluxio tertia nulla eft. Et fic deinceps in infinitum.

Poftquam vero fluentes ex fluxionibus collectæ funt, fi de veritate Conclufionis dubitatur, fluxiones fluentium inventarum viciffim colligendæ funt & cum fluxionibus fub initio propofitis comparandæ. Nam fi prodeunt æquales Conclufio recte fe habet:

Quadr: Tab.I.

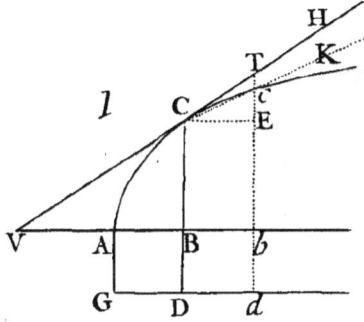

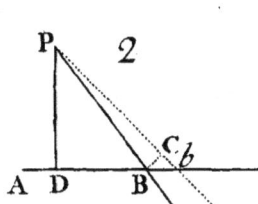

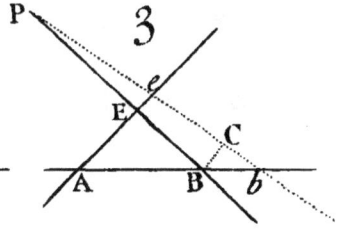

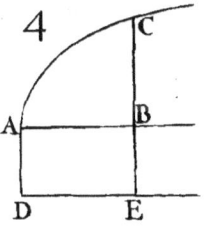

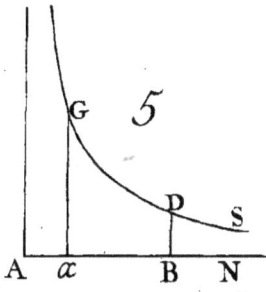

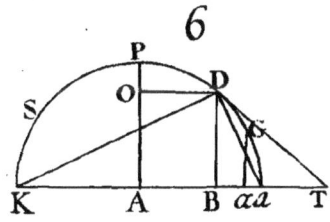

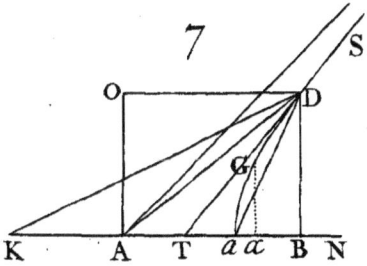

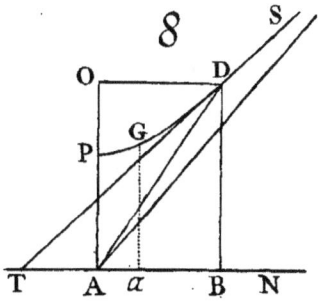

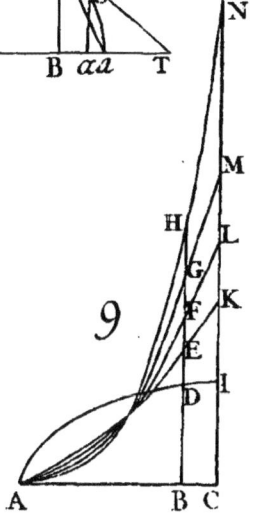

bet : fin minus, corrigendæ funt fluentes fic, ut earum fluxiones fluxionibus fub initio propofitis æquentur. Nam & Fluens pro lubitu affumi poteft & affumptio corrigi ponendo fluxionem fluentis affumptæ æqualem fluxioni propofitæ, & terminos homologos inter fe comparando.

Et his principiis via ad majora fternitur.

F I N I S.

ERRATA.

BOOK I. *Of Opticks.*

PArt 1. p.3. l.20. *Properties which,* ib.p.5. l.5. *and that C,* p.6. l.9. *DE,* p.21. l.23. *are two Rays,* p.27. l.6. *in the Margin put Fig.*14 & 15, p.30. l.7. *MN,* l.9. M, p. 44. l.15. *as was proposed,* p.52. l. 17. *a paper Circle,* p.57. l.ult. *emerging,* p.60. l.25. *contain with the,* p.64. l.18. *and* 14th, p.65. l.13. *at the,* p.66. l.3. *Semicircular,* p.67. l.25. *Center,* l.31. $4\frac{1}{8}$ *Inches,* p.68. l.8. *to* 16, l.9. *or* $5\frac{1}{3}$, p.71. l.1. *bisect,* p.72. l.13. *falls,* l.20. *being.* Part II. p.86. l.5. *lelopipede,* p.89. l.9. *made by,* p.93. l.18. *to* $77\frac{1}{3}$, l.28,29, *by the third Axiom of the first Part of this Book, the Laws,* p.105. l.5. *see repre-sented,* p. 144. l. 24, $\frac{1}{9}, \frac{1}{16}, \frac{1}{10}, \frac{1}{9}, \frac{1}{10}, \frac{1}{16}, \frac{1}{9}$. p. 118, 119. for *Lib.*1. *Lib.*2. write Part I. Part 2. p.122. l.9. *indico,* p. 130. l.19. *to the Angle,* p.132. l.6. *by the bright-ness,* p.135. l.14. For *if in the,* l.16. *first Part you,* p.136. l.26. *first Part,* l.27. *lights,* p.137. l.20. *green, accordingly as,* p. 138. l. 21. *Prop.*6. Part. 2. p.139. l.5. *on which,* p.142. l.17. *XY which have been,* p.143. l.7. *purple,* l.16. *several Lights,* l.24. *of white.*

BOOK. II.

P.5. l. 5. *nicely the,* p. 7. l.9. *y,* ζ *denote,* l.28. *them divers,* p. 10. l. 24. 1000 *to* 1024, p.11. l.11. *oliquities, I.* p. 17. l.4. $14\frac{1}{3}$ *to* 9, p. 25. l. 11. 1 $0\frac{\pm}{3}$, p.31. l. 12. *more com-pounded,* p.55. l.3. *sizes reflect,* l.24. *and therefore their Colours arise,* p.65. l.5. *corpus-cles can,* p.71. l.17. *given breadth,* p.84. l. 4. *are to those,* p. 96. l. 24. *Observation of this Part of this Book,* p.103. l.17. *was to the thickness,* p. 105. l. 19. *of this white Ring,* p.107. l.20. *become equal to the third of those.*

Enumeratio Linearum.

p.143. l.20. *datas signis suis,* p. 144. l.27. *respiciunt,* p. 146. l.5. *sunt Asymptoto,* p. 154. l.13. *cx* $+$ *d dat Ordinatam y* $=$, l.14. *quæ generatur.*

Quadratura Curvarum.

p.168. l.24. *recta AB,* p.176. l.ult. $+ \begin{matrix}\theta\\ \lambda^n\end{matrix}$ *fz^n,* p.183. l.13. *a,b,c,* &c. *e,f,g,* &c. *k,l,m,* &c. p. 185. l.4. *in* zθ-1, p.188. l.14. z$\theta \pm n\sigma$, p. 190. l. 19. *vel* $\overline{x\theta - 1 + 2\lambda\mu}^{\frac{1}{}}$. p. 192. l. 18. g z$^p + 2n$. p.193. l.11. $\overline{aS_v + bR_\tau}^{\omega}$.

www.ingramcontent.com/pod-product-compliance
Lightning Source LLC
Chambersburg PA
CBHW061121220326
41599CB00024B/4119